Spiritual
Theory of Everything

A Unique Blueprint to Discover the Origin and Purpose of Life, Awaken Your Consciousness and Lead a Blissful Life

by
Thomas Vazhakunnathu

thomasvazhakunnathu.com

Table of Contents

Introduction

"You shall know the truth, and the truth shall make you free"

~ Lord Jesus

Galileo, the father of modern science, championed heliocentrism when the whole world around him believed in a geocentric universe. It took a lot of courage for him to stand by what he had discovered. Consequently, he was tried for heresy and put under house arrest until his death. It is true that truth uttered before its time is dangerous for the truth-teller. However, throughout history, great souls revealed truths that the majority of people could not understand at that time and paid for it even with their lives. Because of their sacrifice, those truths were able to bring freedom to subsequent generations when people were able to understand them.

We are fortunate to be living in this digital age. During the last couple of centuries and particularly during the last few decades only, with the rapid development of science and technology, we were able to discover several wonderful truths about the material universe that have freed us from many fears and superstitions. Science and technology have been making our lives easy and comfortable and helping us to unravel more and more mysteries of the universe. The ever-accelerating progress of science and technology made those things possible which were beyond the imagination of our previous generations.

However, even after achieving so much progress in so many areas of life, there are certain questions majority of humanity is still struggling with. These are the fundamental questions about life like 'who we are? what is life? what is the purpose of life? what happens to us after death? who created everything? is there a God? etc. Every human being, at some point in time in his or her life, has stumbled upon these questions unless one has not reached that level of awareness or has decided not to think for himself or herself. Answers to these fundamental questions are called fundamental or universal truths. The purpose of this book is to help answer these fundamental questions and show every person, who is open, his true eternal nature and the oneness of everything in the universe.

Understanding fundamental truths could help us explain most of the phenomena and solve the problems that we face on earth today. Though it is not humanly practical to understand each and every aspect of life and the universe, we could know many truths that are more than sufficient to take care of our current incarnation and live a blissful life. However, no truth can come to us until we are ready for that. We will be able to know truths to the extent of our evolution and seeking only. This is an important concept we have to keep in mind. Therefore, please approach this book with openness and keep in mind that there are always more truths to know, and we will know more as we evolve – as our consciousness or awareness expands (each new information that comes into our awareness expands our awareness and changes the quality of our life).

My journey for truth started with my first awakening experience in 1994. Since then Jesus's words 'the truth shall set you free' (John 8.32) and 'seek and you will find' (Mathew 7.7) kept me on the seeking path. These words made me think whether there is an 'ultimate truth' that can set one free and if such a truth is there, then I wanted to know the same. It has become my passion to seek and know the truth to the extent humanly possible.

For many years I struggled with the question of ultimate truth and searched most of the available resources. After about 10 years of seeking I was able to understand that the 'ultimate truth' for us human beings is that 'All is One and we are eternal spiritual beings experiencing a human life now'. Therefore, the ultimate goal of human life is 'self-realization' - realizing or knowing our true nature - the spiritual nature. It is the salvation, moksha, liberation, nirvana, or enlightenment that humanity is eagerly waiting for.

I also understood that there is a 'truth' in everything, albeit with varying importance, and knowing these truths, specifically the fundamental truths, can free us from many limitations and sufferings and lead to peace, happiness and enlightenment, in an ascending manner and ultimately to a blissful existence. Apart from studying Jesus's messages from the Bible, I studied ancient wisdom teachings (which includes Egyptian, Greek, and Indian (Hindu) philosophies) and channeled materials, particularly Ra materials, which has helped me to know or realize the deeper truths of life and experience higher levels of freedom.

Trying to find 'truth' from one particular religion's scripture is not a wise idea as most of these were revealed to or received/heard by different individuals either directly from the original contributor or from secondary sources, who then memorized them and passed them on orally from one generation to the other before being written down and compiled as books, edited and regularized from time to time and then translated into various languages. During this long and tedious process, there's every possibility that some of the concepts were lost in transmission and some of them were influenced by the writers, editors, and preservers and hence lost their original meaning. Two common mistakes committed by many of these writers were that of elevating the messengers to the level of God or depicting them as God and changing the messages to suit their purpose. I do not want to comment on any specific mistake here but let us not ignore the fact that given the nature of humans and the way these scriptures were originated, preserved, and compiled into books, there are bound to be many things that are not consistent with the teachings of those 'evolved beings' (This is not withstanding the fact that the origin of some scriptures itself is controversial).

Scientific knowledge is updated periodically based on the latest findings, whereas most religions are not ready to even accept that there are errors in their teachings or practices. This is mainly because no religious groups want to accept that parts of their 'holy' book or certain books were edited or are not consistent with the teachings of the messenger for fear of losing their authority. There is no doubt that some religions have more correct and useful teachings compared to others, but there are several incorrect or useless details also, which create confusion and keep the masses in ignorance. However, it will be like throwing the baby out with the bathwater if we totally reject religions and scriptures just because there are mistakes found among them.

It is our duty to seek and find truths with the help of these scriptures and ignore those things that are not in alignment with fundamental/universal truths. If we compare available scriptures, we can see Hindu scriptures (specifically Upanishads and Bhagavad Gita) have detailed information about life, its nature, and purpose. Except for channeled messages (specifically 'Ra Materials' from L/L Research), I could not find such logical and detailed information about life elsewhere. In comparison, many other scriptures deal with life partly or incorrectly. However, fundamental truths are found in every religious book in varying degrees and no particular group or religion can claim exclusivity to them.

Since a large portion of humanity is unable to comprehend the universal truths, it suits them to follow a religious or devotional path until they become ready. As Gautama Buddha said, 'man's ideals must be related to his current state of being', one cannot expect a kindergarten student to work on a college degree. Those who are following/practicing different belief systems should first strive towards achieving the highest ideal of their particular group or religion, so that they will be able/ready to comprehend higher truths. It is also true that there is no point in discussing the ultimate purpose of life or spiritual development with people who are struggling with survival or existential issues. To such people, we need to show compassion and make their life comfortable as far as possible while keeping in mind that there is a reason for everything and most of our experiences are trying to teach us some lessons like love, forgiveness, patience, humility, etc.

Religions offer a support system for people and help them endure pain and suffering. When somebody is under pain or hardship, the immediate objective is to escape from that and, naturally, he or she looks at their god, guru, religious establishments, rituals, and fellowship for necessary support. Religions also offer a kind of social security, at least for people in less developed countries. However, most religions have perpetuated ignorance and kept the masses away from enlightenment. The requirement of religion will go away only when people have evolved beyond those needs. Till then it will be around us in all shapes and sizes.

Traditional Science is correct as far as it has progressed towards the study of the form or the physical side of life and nature. It is always helping us to understand various aspects of the creation. However, even after achieving so much scientific and technological advancements, we collectively still know less than 1% about life and the universe. Don't be surprised when you find many of the truths discussed in this book 'discovered and confirmed' by material scientists in the immediate future. It is worthwhile to note that even with such incomplete information we have made so much material progress, though this partial knowledge has led to the invention of harmful and destructive technologies or misuse of good ones to mindlessly acquire money and control.

No matter how big or useful a discovery, physical science has its limitations and can bring out only part of the truth as only the physical manifestation is subject to such analysis. Though reductive materialism dominates physics, there is hope that quantum mechanics is venturing into the non-physical or metaphysical aspects of reality and has started postulating many theories that look more spiritual than scientific. Until we are able to develop machines that could detect the metaphysical aspects of the universe, we could not scientifically prove how the universe works. Therefore, we need to adopt a multi-disciplinary approach to truth as neither science nor religion can explain it alone. Keep this truth in mind while reading this book so that you will be open to many new possibilities.

This book is the culmination of many years of seeking through spiritual research and experiments to understand life and the universe. It is an attempt to share an overview of the nature and mechanism of life from a spiritual and scientific perspective – a holistic theory of everything. This book covers some of the fundamental matters like:

- God
- Consciousness
- Origin of Life
- Creation
- The Fundamental Particle
- Soul, Mind, and Spirit
- Rays and Chakras
- Cosmic Laws or Natural Laws
- Densities or Dimensions
- Evolution
- Purpose of Life
- Spiritual Development
- Reincarnation
- Karma

- Natural Living
- Planetary Changes

This book will explain that we live in an orderly and purposeful universe and when we learn to live in harmony with It, we can enjoy abundant life in an ascending order. It will also explain how much we can evolve by understanding our true identity/nature and aligning ourselves with the purpose of life. Further, it explains that we are the cause and the conditions of our life are only the effects and when we change inwardly, our world/circumstances must change so as to be is in alignment with our vibrations. This book will be helpful to all seekers of truth, whether scientific or spiritual, and particularly to those who are seeking enlightenment. Many of the concepts need to be dealt with individually to do full justice to them and I hope to take up some of them separately later.

Since my objective is to awaken and help humanity to consciously participate in continuous learning and spiritual evolution rather than establishing a philosophical or religious organization, no attempt has been made to formulate the ideas, methods, and principles presented here into a watertight system. However, a conscious attempt has been made to educate how a person can, by transforming his or her inner nature, live in harmony with Nature and evolve. It is my sincere desire to explain life and the universe using available scientific and spiritual knowledge to bring about much-needed enlightenment to humanity in this new age of transformation. Truths shared in this book will challenge, inspire, transform, and set you free and you will experience love, peace, contentment, and happiness proportionate to your understanding of the truth.

Some of the truths presented herein may not make sense or be acceptable to many at this point in time. However, we have to agree that there can be no real understanding if we listen only to those things that gratify or confirm our existing beliefs and ideas. Unless one seeks truth with an open mind, there is no chance of finding it. Truth will be opened to us to the extent of our openness and seeking nature only. Nobody could prove these things to us; only we will know them when we are ready. It is hard to believe or accept a truth until we know or experience it ourselves. Therefore, be open to the endless possibilities of life and take whatever is plausible to you here and now. Seek and you will find it.

Through this book, I have tried to address important questions of life that are relevant to us now, albeit in an abstract manner. If any of these ideas or concepts are not resonating with you now, then do not constrain yourself and just let it pass. Understanding is a step-by-step process; let it unfold naturally. If something is not relevant or you are not ready at a given point of time, then certain information won't be revealed to you. Therefore, be patient, take your time, take in only what is making sense now, and sincerely try to embody and live it which will, in turn, lead you to higher truths and to self-realization.

I sincerely hope that, together we can consciously evolve and help others and our home planet for a smooth transition.

Thomas Vazhakunnathu

****

Chapter I: Source of Everything

The ultimate source and cause of everything - Its nature and how we can recognize and consciously connect with It. The Oneness of All and the purpose of life.

Introduction

Since ‘nothing’ cannot create or become something, there should be 'something' to create or become something else. This principle applies to everything including the creation of the universe. As per modern science, everything originated from a Singularity, a point of infinite density, and there was an explosion called 'Big Bang' which lead to the creation of this universe. Bible says, 'In the beginning was the Word, and the Word was with God, and the Word was God' (John 1:1). We can see that the Greek word translated as ‘Word’ is ‘Logos’ which means a creative power or principle or an aspect of God. Some consider this ‘Word’ as a sound vibration and some as a thought vibration. As per the Hindu scriptures, everything originated from a formless being and Om was the beginning of creation. Om is the sound caused by certain vibration. Either way, there should be someone or something to create a sound or a thought. Though the concept of God varies from religion to religion, every religion agrees that God has created everything. So, we can see, as per both science and religion, there was something or someone before the manifestation of the universe. We may call this ‘Source and Cause of Everything’ as Singularity, Spirit, God, Infinite intelligence, The Universal, *Paramatma,* Supreme Soul, The All, Alpha and Omega, or whatever name we have heard or thought about the Creator. I prefer to call this source and cause of everything as 'God'.

Everyone will agree that a lot of intelligence, resources, and hard work are required for any big project on this planet. Now, imagine, how much intelligence, resources, and efforts are needed to create and maintain this vast universe. Even with a basic understanding of the intelligent design, magnitude, and complexity of the universe, it can be easily concluded that this is the work of an infinitely intelligent and omnipotent Being. Through this section of the book, let us try to understand this Being - the Ultimate or Supreme Reality of the Universe. When we have established that the source and cause of everything can be traced to one ultimate reality, then we have a strong foundation to build our philosophy of life.

There are many scientists and philosophers who believe that the source and cause of everything is some kind of 'Intelligent Energy' which is beyond the scrutiny of the current tools of science. However, particle physicists still hold the reductionist view that everything can be broken down into fundamental particles and they continue to search in the effects for the cause. World's largest and most powerful particle accelerator as of now, the Large Hadron Collider, has been smashing atoms for the past several years but is yet to find that elusive particle which is the source and cause of everything. The truth is that if we assume the smallest identifiable portion of 'matter or energy' as the 'source and cause' behind this intelligent and purposeful universe, there is no way to find the same. What creates these energies and how they are being created and evolved is beyond physical scrutiny, as the controlling principle and power, the intelligent life, within these physical levels cannot be observed or measured by the tools of science available to us now.

Newton's Laws of Motion and Einstein's General Theory of Relativity explain how the universe works at a macro level and quantum mechanics explains it at the subatomic level. Similarly, Darwin's theory of evolution explains the biological evolution of species, basically the evolution of their bodies. However, these scientific models are inadequate to explain or understand even the fundamental nature of the physical universe, let alone life. Human limitations and lack of technology are not allowing us to observe and study the metaphysical universe. It does not mean that such a reality does not exist. This only means, we do not have the capability or instrumentation to verify it now and may be able to do so as and when we are equipped.

However, it is worthwhile to note that scientists like Newton and Einstein used their mental faculties more than the physical tools for their search for fundamental truths and got such astounding results which still keep people wondering how they achieved it. Further, there have been many evolved individuals throughout the history of mankind who were able to see and act beyond the material world and shared those higher truths for the benefit of humanity, whose wisdom is partly recorded in some of the ancient scriptures. It is, therefore, logical to conclude that, more than the mechanical means, mental faculties are better suited for investigating God, the 'Source and Cause of Everything'.

Nature of God

Currently, most religions hold a distorted view of God, though many of their scriptures describe it as an absolute, omnipresent, omnipotent, omniscient, conscious, and intelligent universal Being, creating everything out of Itself and yet remaining unaffected amidst the ever-changing creation. All these descriptions are true to the reality of God, who is the infinite intelligence and intelligent energy that fills the universe, whose presence is felt on the spiritual realm as consciousness, on the mental level as thoughts, and on the physical level as matter. The fundamental nature of consciousness is love, which is unconditional, unbiased, undistorted, and does not discriminate, differentiate, or change according to circumstances. The intense feeling of deep affection we experience in our life is an expression of this love.

If we visualize God as a large sphere of white light, then the universe consisting of numerous galaxies could be considered as a small sphere of glass situated at the center of this light. God is the eternal space (all the space), from/in which everything manifests as the physical and metaphysical universe. It is the singularity or unity that is in motion, the power that is moving the galaxies and everything in it – physically and spiritually. It is the alpha and omega and nothing can be added to or subtracted from it. It cannot be confined to any form, shape, or locality as it incorporates all things and there is nothing outside of it.

As of now, it has been estimated that our Sun's mass consists of about 99.9% of the total mass of the solar system. The remaining 0.1% is shared by all the planets and other celestial bodies. Out of this 0.1%, 90% of the mass is shared by just two planets - Jupiter and Saturn. The mass of the earth is only about 0.0003% of the total mass of this solar system, whereas the Milky Way galaxy, our home galaxy, consists of roughly 250 billion stars (suns), and most of them have several planets around them. Among these planets, there are millions of planets that could support biological life like our planet earth. Since the number of galaxies could not be even estimated, the size of the Universe is beyond the perception of the human mind and then to think that it can only exist within God is an incomprehensible subject to many.

Many people are obsessed with the material world and have become so conditioned and deep-rooted in their earthly existence that they still tend to perceive the earth and human beings as the center of the universe and life, and this misconception is accentuated and perpetuated by some mainstream religions. Now with the advancement in science and technology, such false beliefs have been exposed and those who have acquired some knowledge about the universe know our relative position within the universe.

Though there have been attempts by certain Governments and organizations to underplay, misinform, and surreptitiously put under the carpet evidence of UFOs and ET life, results of recent space research programs suggest that humanity is not alone in this universe. Even from a probabilistic point of view, mathematically, and taking into account the size of the universe and the scientific progress we are making, one should not be surprised to hear about life similar to us on other planets or exo-planets (planets outside the solar system) through the mainstream media soon. Therefore, we should be humble enough to accept that there will be many areas of God that will remain hidden to us while in a physical body and at this level of awareness and that there is much-much more in the universe that we cannot see than we see now.

Many people have been taught that God is a personal being who rules the world just like a benevolent King (in some cultures God is even limited to certain groups of people, villages, or regions). People have also been taught that there is another powerful, but, malevolent King (Devil), whose only job is to fight with God. People, who are yet to become self-aware, continue to believe such concepts because they could not imagine any other possibility. The truth is that there is no duality in God or Its primary creation. However, as the creation begins experiencing life, everything appears with its pair of opposites, *Yin -Yang* nature, which is the 'Divine Paradox'. This *Yin* and

Yang natures are actually complementary and not opposing ones. They are the two sides of the same coin and remain inseparable. We can see this nature everywhere in the universe at every level of experience, like physical and metaphysical, positive and negative, matter and antimatter, female and male, day and night, light and dark, hot and cold, liquid and solid, inhale and exhale, pleasure and pain, love and hate, fear and courage, good and evil, birth and death, creation and destruction, etc.

An atom may be negative not because it does not have a positive nucleus but because of a change in its constitution as the number of electrons are exceeding the number of its protons. This means that there is positive and negative within an atom which is true to every aspect of the creation. Opposites not only help to preserve the balance or equilibrium in Nature but also provide life its much-needed contrast. One can understand the true value of anything only when it is seen in a contrast. If there were no opposites, then there is nothing to contrast and hence no scope for experience and consequent learning as there is hardly any impetus to do anything. Therefore, this *Yin-Yang* nature of creation is necessary for the balanced evolution of the universe as it guides and helps the soul to make a choice, experience the consequence, and learn the lesson. A closer look reveals that there is no discrete boundary between these seemingly opposing natures and they gradually shade into each other.

God being the ultimate power and presence of the universe, nothing can challenge Its authority and there cannot be a competitor to It. However, the concepts of 'All is One' and 'God is Love' do not deny the existence of evil or negative souls. There exist negative souls, in a hierarchical nature within the creation beyond human existence, just like there are negative people among us. These are the souls who chose the negative path during their human existence. Maybe, to explain these negative souls and their influence, some religions have invented the concept of 'Devil'.

God does not belong to any religion because it has nothing to do with religion. God is of its own accord and does not require the assistance or approval of human beings to exercise its absolute power and presence, though some fanatics think they have to protect their helpless God. For that matter, even humans cannot claim any exclusivity to God because we and our planet is nothing but a drop in the ocean of this endless creation.

The Universe – the galaxies and everything in it - works based on immutable Laws which we identify as the Cosmic or Natural or Universal or Fundamental Laws. Some are laws of physics and some are metaphysical, both having the same accuracy or precision, which are the natural tendencies of life or creation as inherited from God and they govern all the interactions among all creation. These Laws ensure that the order and balance or rhythm of the universe is always maintained.

We are the individualized manifestation of God, experiencing life using a physical/ chemical body of human form at this point in time in our evolutionary journey. At the existential level, all our lives are inter-connected and inter-dependent as we share the same consciousness and its distortions and manifestations such as mind, energy, breath, space, and environment. We can compare our existence as an individual portion of God to one of the trillions of cells that constitute our physical body. Though all the cells have a certain level of independence, they are dependent on each other and subservient to the controlling intelligence of our body under the control of mind and work towards a common goal.

Whether we consciously realize or not we are always one with God and Its creations. Apart from ancient religions that are extinct, Hinduism* and its offshoots, and mystics of Abrahamic religions believe that 'All is One'. Jesus and Krishna taught that 'All is One' and the importance of achieving a conscious understanding and experience of 'Union' (yoga) or 'Oneness 'with God. Anyone who cares about truth and is ready to sincerely seek will surely find and experience this union in an ascending order.

Note: *Hindu *Sanatana Dharma* is a philosophy based on 'All is One'

How to know God

We are living in a multiverse consisting of countless galaxies and individual beings or life units existing in multiple densities or dimensions. There is a hierarchical structure and function in the universe as well as within the individual and the higher always incorporates and guides the lower levels. Every life unit consciously functions in a particular physical vibratory level corresponding to its stage in development while in the incarnation, but subconsciously functions in and through the metaphysical levels which interpenetrate the physical universe, and, as it evolves, moves to the next vibratory level. We, as human beings, consciously function in the material world of planet earth with the help of a physical or chemical body but subconsciously function in the metaphysical levels using our metaphysical body (soul). At death, our soul leaves the physical body and continues to function in the same universal field, but on a metaphysical vibratory level till the next physical incarnation in case of continuing human life or transitions to a higher dimension of existence, if the requisite development has been attained.

God's basic nature is love and we, as individualized portions of God, are not any different. We feel at peace only when we can maintain that natural state. Anything that is void of love causes conflict and restlessness. In our ignorance, we think we are separate from others and mistreat each other, which we all know is wrong, deep within us. Ignorance of self (not knowing who we are or our true nature) and consequent disconnection from other-selves and Nature is the source of all problems we are facing in this world.

We are individually and collectively responsible for the reality that we create and experience on this beautiful planet. There are enough resources for everyone to live a happy life. However, the majority of people are struggling to survive because we don't know how to live. The scarcity and hardships that many have to endure are because we don't care for each other. Whereas the principle to be applied is 'do to others what we expect them to do to us' because All is One and we reap what we sow.

Instead of addressing this 'ignorance of self', religions try to make life comfortable or tolerable by ignoring or distracting or focusing on material comforts, whereas governments handle it through punitive systems demanding conformity and obedience to its rules. Ignorance of self is the biggest enemy of humanity. Let's fight it out rather than creating divisions and fighting with our fellow beings and nature.

The ultimate purpose of life is to know or become aware of the self. In other words, gaining self-knowledge or self-realization is the purpose of life. To the extent we understand ourselves or our true nature, to that extent, we can understand other selves, the universe in general, and the Creator or

God as fundamentally 'All is One'. We can never experience the peace that surpasses all understanding until we realize who we really are. This realization and consequent experiences are possible proportionate to our spiritual development, which is also known as soul or consciousness or self or inner development. In other words, knowing God is our journey of 'self-realization', an eternal journey that ends with our complete union with God.

Knowing begins with an inquiry or a seeking. When we try to investigate our world, we find that the deeper we go the more it becomes evident that there is some power and presence behind everything seen (physical) and unseen (metaphysical). For example, let us examine the functioning of our physical body where each organ does its assigned work without our conscious knowledge. When we sleep, we lose the conscious control of the physical body, but the body is busy doing so many important functions while we live our dream life. Who takes care of these activities? Now consider our planet, solar system, and the galaxies. How do they operate so perfectly? Who maintains the balance and harmony of this universe?

As we progress in this investigation, we will slowly begin to understand the intelligent, purposeful, and reciprocal nature of everything seen and unseen in the universe. We will also realize that the outer visible world is the expression or physical manifestation of an invisible world which is metaphysical. We will also understand that everything is interconnected, and All is One and start experiencing this Oneness with the Creator and its creations in an ascending order, to the extent possible as a human being. Every sincere seeker will know and experience God in direct proportion to the spiritual progress he has attained.

As a person becomes awake (comes out of the conditioned mind), there comes a realization that he is not a 'physical being' but a 'soul, a 'metaphysical being', presently experiencing life as a human being with the help of a physical body. This realization will cause him to look within instead of searching outside and find the power and presence of his soul. This is an important step of awakening, the beginning of our conscious evolution or self-realization - a process lasting the entire life span of a soul and realized progressively in stages. Of course, there could be moments of profound experiences, but they are just consequences of the sudden expansion of consciousness or realization of truth which happens many times to a conscious soul while inhabiting a physical body.

Soul, our eternal self or true nature, is not visible to our physical eyes but can be experienced as a presence and understood/ seen with the mind when we direct our attention inward. The more we understand this inner nature, the closer we get to God. Until a person is free from the superstitious beliefs of a human-like god or gods ruling over us in an arbitrary manner, it is not really possible to make any meaningful spiritual progress. It is those who do not know God who make it an object of worship. Because people tend to worship what they fear (to escape the wrath), what they presume as powerful (to get favors) and what is beyond their reach (hero-worship – a self-gratification method mostly used by people who lack self-esteem/ confidence).

Most people are unable to comprehend this truth easily because our physical sense organs are equipped to handle only the physical world, which is the outermost plane of manifestation. It is our identification with the physical world and its cares, especially our physical body with its sensual nature that prevents us from knowing our true nature and oneness with God. The more a person gets detached from the physical world, the easier it is for him to work in the metaphysical world. Therefore, those who struggle with their physical body and circumstances must first get them under control to make any serious investigation about the soul (Yoga system was developed precisely for this purpose). Our focus should be on removing the obstacles to our spiritual development and to arrange our environment in such a way that our mind can maintain a calm and steady state continuously.

Once we consciously enter the path of spirituality, there comes a sense of purpose, security, freedom, and abundance in life. We will be able to consciously co-create with nature whatever we choose. Life becomes peaceful, happy and easy as we always find ourselves being helped and supported by both physical and metaphysical worlds. Even when there is chaos all around, we will be able to maintain calmness and remain unaffected. And we are better equipped to manage painful or difficult situations, smoothly and gracefully because the mind is calm and peaceful.

When we are deep into spirituality then we will be able to consciously experience the divine/source energy connection, protection, love, and provision continuously and we will never feel insecure or lonely. We will know and experience that all things work together for our good. As we progress, we become more connected and experience a blissful life. This is the reason why great masters were able to live alone in the wilderness. India had many such sages who spent their entire life alone in the inhospitable terrains of the Himalayas. Some of them are known to have attained great wisdom and supernatural powers and contributed much to the development of humanity.

Knowing or realizing fundamental truths expands the consciousness and we get a larger perspective, where all other things fall into place and start making sense. Slowly we will be able to understand the true nature of everything and accept and appreciate their relative value and purpose. We should, therefore, focus on those things that help us to progress spiritually and ignore or give less priority to everything else. If you are already a conscious seeker then avoid belief systems/ programs and 'human gods' promising enlightenment as they may enslave you and prevent you from further learning during the current incarnation. Don't be dependent on anyone or anything other than yourself and God. Be always open to change and make physical activities and meditation part of your daily routine. Start trusting and using intuition and dreams more because they are powerful tools available for investigating the metaphysical reality and learning. We should always try to live in harmony with Nature and let the awakening or expansion of consciousness happen naturally and smoothly like a sapling growing into a tree and bringing forth beautiful flowers and fruits.

Those who already know who they are can easily analyze their thoughts and make necessary changes and consciously make strong progress in this journey of self-realization. Those who have not yet reached that level may have to depend on spiritual teachers and sincerely try to learn from them till they understand who they are. No higher truths can come to us until we are ready, as our beliefs determine the boundaries of our perception and imagination, and we can be conscious and experience reality to that extent only. In other words, we can be conscious of only what we can visualize - first we have to imagine the possibility of something to become conscious about it. Since learning is a step by step process, we start at a lower level and move higher as we progress. This is the reason why every individual has to go through different dimensions and numerous incarnations to complete their self-realization journey.

We seek truth eternally. As a human being, we seek truths about earthly life and the life hereafter. When we transcend human life then the quest for the truth will be knowing that new life and life thereafter and so on. When we seek, revelation (unveiling) of truth comes proportionate to our commitment and readiness. There is no way for our limited intelligence to figure out all the secrets of God or the universe at one go - we realize in stages as we evolve. Those who sincerely seek will surely find and reach higher levels of understanding.

> ***Note:*** *More about the soul, spiritual development, and living in harmony with Nature are discussed in the next few chapters.*

****

Chapter II: Involution, The Creation

Creation of the fundamental particle, the soul and manifestation of the Central Sun, the Galactic and Solar Systems

---

Introduction

Many believe that an event called the 'Big Bang' caused creation of the universe. Despite being a popular theory among the scientific community and intellectuals, it appears to be just another attempt to explain the origin of the universe from a material perspective. Though there is no possibility of the Big Bang having caused the initial creation, it is an easy way of explaining the initial creation like saying 'in the beginning, God created the heavens and the earth'. Major lacunas in the theory tell us that it was introduced as a fill-gap arrangement to explain the creation in the light of other theories like redshift (an expanding universe), and general relativity. Despite the popularity of the theory, there are many scientists who seriously think about what could have happened before the big bang and try to explain it using various other theories. The following reasons are more than enough to prove that the Big Bang theory is not the event that caused the initial/ original creation:

a) An explosion or rapid expansion is impossible to happen out of nowhere as something material had to exist before such an event to happen.
b) As per the Standard Model of particle physics, the smallest particles join together to form subatomic particles which then form an atom and atoms form a molecule, and so on. When this is the mechanism for

the matter to evolve and accumulate mass then what could have caused an explosion before the manifestation of even the fundamental particle?

c) Also, as per metaphysical science, everything begins metaphysically and attains materialization after a considerable amount of time, may be billions of years in the case of initial creation. The metaphysical activity does not cause any kind of explosion as there is no physical matter/energy involved that can give rise to such a phenomenon.
d) The study of black holes and the birth and death of galaxies tell us that the big bang theory is no more needed to explain the initial creation.

However, we cannot entirely discredit the Big Bang theory, as it appears to be one of the events that happened during the chain of events in the creation process. Though the Big Bang theory can be considered a baby step towards understanding this infinite and mysterious universe, it has effectively denied the existence of an intelligent and omnipotent Creator. Big Bang theory is a classic example of the reductionist view of a purely materialistic and mechanical approach that considers everything as material with limited use, value, or purpose.

Even a casual glance at our immediate environment reveals the wonderful nature of this magnificent universe and any further investigation proves beyond doubt that there is a great amount of Intelligence at work behind the intelligent design and plan of every part of the creation starting from the fundamental particles to the galaxies in multiple densities or dimensions. Now, to say that the universe originated from nothing and that there is no purpose in it is the height of ignorance.

When we understand the nature of God, which is infinitely intelligent and omnipotent, we don't really have to look for some explosion to explain the force behind the swirling and expanding nature of the universe. Through this chapter, I am trying to explain how the one God became numerous individual units and manifested as the universe. Involution is the eternal creative process of God or Super soul involving or investing Itself as souls and manifesting as the universe.

Soul - The Primary Creation

In the beginning, there was nothing but God in its undistorted form which can be considered as infinite consciousness or love, appearing as the eternal space, whose presence could be felt like unknown or dark or subtle energy. At some point in eternity, God or Super soul thought or imagined or planned the concept of many-ness with unlimited possibilities of experience and then initiated the process of becoming numerous individual souls.

This exploration began as a small disturbance or movement in the eternal sea of subtle energy. At the initial stages of creation, there appeared vortexes or foci of angular movements in this subtle energy, with free will (the force aspect of consciousness) acting as the catalyst of this movement. Over a long period of time, the speed of these angular movements gained momentum, and having reached high velocity caused the formation or creation of individual souls. This is considered as the action of repulsion of Super soul whereby primary creation happens.

The soul is the smallest individualized portion of God and a co-creator with It. It is an individualized complex of consciousness or a quantum field of consciousness, a 'soul' with the idea of separate existence. Though each soul is independent and has free will to choose, they are connected and dependent upon each other and the Super soul because of the underlying unity or oneness. The blueprint for development is inherent to the soul, which is the same for all the souls and hence will develop or evolve in such and such a way within a particular environment.

The initial stages of creation were metaphysical and beyond mechanical scrutiny until light or photon was formed. The metaphysical realm of existence is not only subtle but also fundamental, and it is out of the metaphysical that the physical comes into being. The existence and experience of space-time (physical level – both visible and invisible to humans at this point of time) comes into being only after the individuation process of the galactic central systems has been completed and physical forms emerge. Therefore, it is important to consider the time-space (metaphysical level) and space-time aspects of the universe and the creative process while studying these fundamental truths. What we could observe/analyze/estimate with our current technology/ instruments are only the forms, their physical nature and age after the creation has entered into space-time.

Souls manifesting as Photons

After a long period of soul's interaction with other souls and the Super soul, under extreme conditions of a black hole, a field or body of electromagnetic energy was formed and manifested as light or photon. Photon or light is a soul manifesting with a light body, the primary manifestation of a soul. It is a discrete quantum of energy or an intelligent energy complex or an electromagnetic wave, which is the fundamental particle of the universe.

The vortexes of photons manifested as the central suns or central systems which caused enough chaos leading to the creation of galaxies with billions of solar systems consisting of suns, planets, and other celestial bodies over a long period of time.

Each central sun or galactic sun is a primary co-creator with the Infinite Creator or God. This co-creator has the free will to determine the nature of photon/intelligent energy which promotes the lessons within its galaxy. Therefore, the fundamental characteristics of photons slightly vary from galaxy to galaxy. Our sun is the local representative of the central sun of the Milky Way galaxy, that co-created and governs the evolution of the solar system including our planet.

A pictorial representation of a photon, depicting its constitution, is given below:

(Don't bother about the shape; it is only for the sake of understanding, not an exact representation):

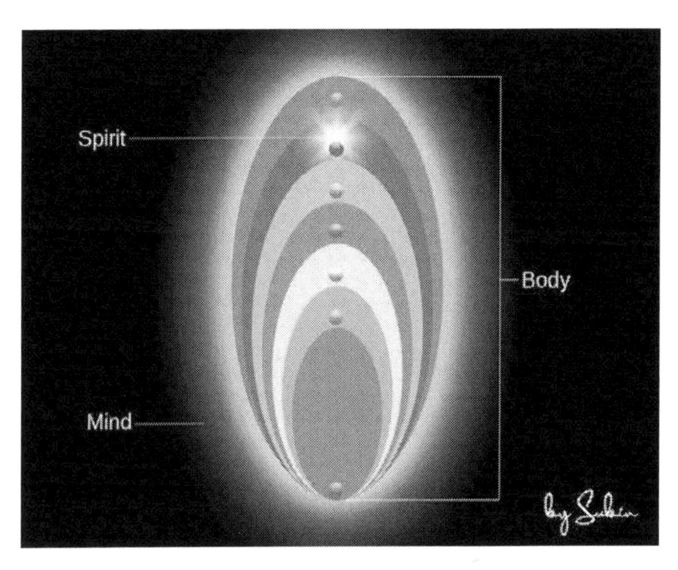

Photon is a web of energy fields or complexes – having three fundamental hierarchical complexes namely the light body, mind, and spirit, as detailed below:

1) **Light body** is the energy complex created by the soul as a body, consisting primarily of seven layers called 'Rays', each layer corresponding to a particular density of existence with a unique frequency and other properties. Each Ray can be identified with one of the seven colors of white light. These rays are the source and blueprint of every physical and metaphysical manifestation of the universe.

2) **Mind** is that energy complex which connects the body complex with the spirit complex and makes the soul function as an individual unit. The mind complex is the communication center of the soul.

3) **Spirit** is that energy complex which is the co-creator within the individual soul and acts as a channel between the individual soul and the Universal or Super soul and other souls. This portion consists of the least distorted energy of the Creator.

The body, mind, and spirit complexes are inextricably joined together, and one complex cannot function without the other. Therefore, the basic constitution of all photons is the same.

Souls initially appear individually as photons and collectively as a photon field or *prana*, the invisible field that permeates all of space, and help each other to evolve. Photons display wave-particle duality and their further evolution (appears at varying speeds and frequencies) causes or creates the matter and antimatter and manifests as the two fundamental aspects or sectors of the universe – the material or the physical (tangible) and the cosmic or astral or metaphysical (intangible).

As per the Standard Model, the quantum field, which is called Higgs Field, is supposed to be responsible for giving particles their masses and the particle associated with the Higgs field is called the Higgs boson. Scientists observed that the Higgs boson decays into a pair of photons along with few pairs of other particles, which also decay into photons. Further, other elementary particles also decay into photons. Since all elementary particles except photon decay into photons and photons do not decay, we can safely conclude that photon is the ultimate elementary or fundamental particle which join together to form particles that constitute subatomic particles and that the fundamental quantum field is the photon field. A photon may have less mass but are not massless – how can something have zero mass and still exist? It appears massless only because of our inability to measure.

The universe was formed when photons, under the direction of spirit (which is under the influence of the co-creators above it and the Super soul), started attracting each other and formed subatomic particles, atoms, and molecules and manifested as ether, gas, fire, liquid and solid forms/states. This is considered as the great action of attraction by the Super soul, whereby secondary creation happens. Every form of creation, from that of a subatomic particle to a galaxy, has an evolved photon as the controlling soul. Since there is nothing but souls in the universe, everything is living and only the degree of livingness varies according to the stage of its evolution.

Einstein discovered that matter and energy are one, just two sides of the same coin, and introduced the famous equation $E=MC^2$, where M is mass, E is energy, and C is the speed of light in a vacuum. Matter can be energy and energy can be matter – completely interchangeable – only the forms and states vary. Now, what is this energy? It is light or photons in its various forms. Photon condenses (by slowing the vibrations) into subatomic particles and forms atoms, molecules, cells, and all other forms, physical as well as metaphysical, in the universe. So, matter is simply energy at a comparatively slow vibration.

As per quantum mechanics, there is no reality until it is observed. This means, without an observer, energy does not take any form. Who is this observer? It is the soul as the co-creator. In other words, soul acts upon energy and creates different forms and states in an ascending order. As the whole creation consists of photons, we can see all the properties of the light body, mind, and spirit in everything in varying degrees according to the level of development it has achieved. The fundamental forces of the nature are the properties of the soul/photon. Forces and fields do not have an independent existence but are the product of the interaction of waves and particles.

The universe at the fundamental level consists of photons vibrating at different frequencies in a delicate and dynamic interplay. As the photon evolves, it is no longer considered a photon but a soul that uses a body to experience life and evolve further. This body may be of a material or subtle nature depending upon the dimension and the incarnation it takes. Therefore, the word photon is used in this book to denote the primary manifestation of a soul as a free particle or a fundamental particle only. Souls do not have fixed size but a flexible nature and adjust according to the size of the subtle or material bodies they inhabit. The core vibrational frequency determines its particular density and locus at any given point in time. Souls, after billions of years of evolution, have reached our current stage, that of human beings. Soul is our true/ core / fundamental nature - the spiritual being or the metaphysical being that experiences life using a physical/ chemical body. We will discuss how life evolved on earth, in the next chapter titled 'Evolution'.

Every celestial body exists in multiple densities as well as sub densities (also known as inner planes), which is true for the earth also. As per the Hindu philosophy, these densities/sub-densities are called *Lokas* (which means worlds) and there are 14 of them - seven higher (densities) and seven lower ones (sub-densities). Some of the sub-densities are considered as 'light' (heaven) and some as 'dark' (hell), depending on their vibrational frequency. The densities are for the space-time experience of the souls whereas sub-densities are the resting/ healing place of souls in between incarnations, where they have a time-space existence without a physical body. All these dimensions (densities) and sub-dimensions (sub-densities) have their unique structure and frequency and are governed by the Natural Laws.

The physical dimensions that we can perceive with our five senses are some of the dimensions of existence. Other dimensions and sub-dimensions of this multi-dimensional universe are intangible in the sense that they are beyond the perceptional ability of humans, though they are filled with life that can see us and influence us. These dimensions and sub-dimensions can be compared to the existence of various types of electromagnetic radiations such as gamma rays, x-rays, ultraviolet rays, white light, infrared rays, microwave, radio waves, etc. within the atmosphere of planet earth. Though most of these radiations cannot be seen with naked eyes, we can see and use them for various purposes with the help of appropriate tools. All these radiations coexist and interpenetrate each other without losing their individual identity. Even the same type of radiations does not interfere with each other as long as their frequency is kept different.

With the help of quantum mechanics, it has now been confirmed that there exists an unseen side in the world and it is unseen only because of the limitations of our senses and the available instrumentation. As scientific knowledge and technology advances, we will be able to verify more of these mysteries of the universe. Though traditional science has progressed deep into the material world, it hardly even recognizes the metaphysical world, which is far more stable and certain than the material world. This may be because there were pressing material needs as well as business opportunities that caught the attention of individuals, corporations, and governments, and thus such researches got enough support and funding. However, this one-sided approach has led to serious environmental and social challenges to our planet and its inhabitants.

Irrespective of the stages/ dimensions or forms, the soul/photon contains all the fundamental characteristics of God, proportionate to the consciousness it holds, as it is an individualized spark/ portion of God. Soul appears sleeping (without movement) as minerals, awake (aware) and living (moving and growing) as plants and animals, self-conscious (self-aware) as humans, and experiencing cosmic consciousness as angels in an ascending order till it merges with God. As the soul evolves, it can process/ use more of *prana*. Just like air is essential for the physical body to survive, *prana* is essential for the soul.

Souls exist for the purpose of experiencing life in unity with God in an ascending order for a period lasting a few billion years (considered as a major cycle) before being returned to or merged with the Creator, in an event called the great dissolution or *maha pralaya* after which another major cycle begins. As the soul evolves, its capacity to experience oneness with God increases and finally attains complete oneness or unity and merges with it.

Note: *High-speed movement/rotation can produce a booming sound like 'Oom or Aum'. Is it a coincidence that the Vedas says, 'the creation emerged from Aum'? Whereas the Bible says, 'in the beginning, there was Word' (is not this Word 'Aum'?).*

****

Chapter III: Evolution, the Eternal Journey of the Soul

The process of evolution of souls and the universes. Tracing each major stage of our eternal journey individually and collectively.

Introduction

In this chapter, we are going to trace the journey of souls through various levels of existence, evolving in stages and reaching the level of humans and then higher beings in ascending order before returning to the initial formless state.

Evolution is the eternal journey of souls through which the Creator knows Itself as creation. Evolution explains the method and process through which souls individually and collectively learn, grow, and eventually return to the Creator. Evolution is primarily about the growth and development of the soul, both in terms of quantum and quality, and secondarily about the co-creators managing the process of evolution. Qualitative development is the tuning or refinement of the consciousness, whereas the quantitative development consists of an increase in the amount of energy of the soul. In other words, evolution is the eternal 'growth and refining' or 'transformation' of the soul in ascending order.

Evolution of the physical or chemical body is secondary, as once the appropriate physical forms or bodies are customized and adapted or the path is made for each stage, there is practically no reason for the evolution of the body from one species to another. Further, every species of organism continues to exist until their intended purpose is served or the environment is capable of supporting them. Physical-world and its evolution is caused by and for the evolution of the souls only.

Therefore, now we do not see an ape body transforming into a human body as human souls no more incarnate in ape bodies.

Traditional science, based on the theory of organic evolution as postulated by Charles Darwin, sees evolution as a consequence of adaptive changes to the environment by various forms of physical life, through genetic mutation and natural selection over a long period, which is correct as far as the process of evolution is concerned. However, this theory is not adequate to explain the purpose of either evolution or life and the reason why an ape body is no more transforming into a human body. This is where the relevance of spiritual science comes, which explains evolution as the eternal journey of souls with the help of numerous physical vehicles appropriate to its stage in development.

Everything in the universe is in a flux and nothing can or is allowed to stand still. And nothing goes through the evolutionary process alone or without impacting others. Evolution causes all manifested life (individually and collectively) to spiral upward towards the oneness or purest vibration of God. Evolution is caused by the ever-increasing urge of the purest intelligent energy (spirit) within the soul to express and experience in ascending degrees. Evolution explains the working nature/ mechanism of the manifested universe and life. Evolution is what we are most concerned about because that is the reason behind our incarnation.

The Journey of Evolution

Through the evolutionary process of billions of years, the universe with all its stars and planets has evolved, which, after a certain period of further evolution, will disintegrate and disappear one after another and then another cycle of creation will begin afresh. All the Suns at the end of their useful lives become nebulas, after collecting all the planets and other celestial bodies (both physical and metaphysical portions) within their system, and then join a black hole, and, after a certain time, take rebirth as new systems. Logically the size of black holes should be more than the size of the physical universe as that is the womb from where everything manifests and also returns. It appears that the entire universe is not dissolved and recreated in one go but happens at the galactic level based on the age and location of the system. At any point in time, there could be several galaxies under dissolution and recreation. Everything in the universe happens in a cyclical manner, which includes creation and dissolution.

What we are discussing in this chapter is the evolution of souls in one such major cycle of creation and dissolution that spans a few billion years, with particular reference to our current life on planet earth. Soul's evolution in this major cycle is broadly classified into seven densities or dimensions or stages or realms or *lokas* or worlds, each of which corresponds to a unique level of vibrational frequency and awareness. The seven stages of evolution are:

1) Simple awareness
2) Growing awareness
3) Self-awareness
4) Love
5) Wisdom
6) Balance
7) Unity or Oneness

These are the main stages of our life at a macro level compared to the seven stages of life at a micro-level during one incarnation viz. prenatal, infancy, childhood, adolescence, adulthood, middle age, and old age.

Apart from these seven stages of evolution, there is one more stage considered as the eighth, which is the state of God as a co-creator. This is the density from which the creation emerges (as the first density) and at the end of a major cycle merges with the Creator (towards the end of the seventh density). The black holes represent/are part of the eight density where the physical manifestation starts and ends in a cyclical manner, eternally.

The nature of the unique space-time manifestation of each density is based on the configurations (the blueprint) placed upon the fundamental particles or photons at the time of their creation. Each density operates within a certain range of consciousness or awareness and accordingly the manifestations and experiences vary for each density. Graduation or promotion or transition from a lower density to a higher density is based on the development of the soul to a particular level which manifests as a certain pattern and frequency in the light body of the soul.

As per spiritual science, planet Earth has evolved to the level of third density and coexists with first and second densities. It is now transforming to fourth density and hence slowly becoming unhabitable for humans and some of the plant and animal life. Whereas, our neighbor Venus is a sixth-density planet and Mars a third-density one. Mars is currently unable to support second and third-density life, except maybe for some unicellular organisms, due to the destruction of its atmosphere. Similarly, each planet within our solar system is at a particular stage or density and is progressing towards its next higher level, which is true of all planets throughout the universe. The physical forms visible to us as humans belong to the 1st, 2nd, and 3rd density manifestations.

Souls learn lessons through experience and knowledge, through which they understand the truth and evolve. How long it takes to learn those lessons determines how long they remain in a particular density of existence. Each soul has the potential to evolve through all the densities and there is no way to force (religiously or otherwise) an entity to move up the density ladder until it is ready and chooses to do so. However, it appears that only a fraction of the souls gets the opportunity to evolve through all the densities in any cycle of creation.

If we compare the planets within our solar system, we could see Venus and Mars evolved faster than Earth whereas Mars has fallen below Earth now. Other planets are progressing at different levels. Also, there will be some planets that would not progress beyond first density by the end of the major cycle and will get recycled when our solar system or our galaxy itself is eventually consumed by a black hole. Similarly, within a planet also, all the souls do not progress linearly as all of them do not get a chance to evolve through all the densities in any given major cycle. Some grow faster, some slower and some get stuck. There are several factors, both internal and external, that contribute to the speed of evolution of souls, many of which are beyond the comprehension of our limited understanding.

We as human beings have got the privilege of reaching the level of self-awareness (third density) and hence have a better chance of reaching higher levels before our eventual merger with God. We can consider ourselves as the chosen and privileged children of God (compared to mineral, plant, and animal life), destined for higher levels of existence, even though earlier we had no clue of what was happening to us and were utterly lost in the darkness of ignorance. It can be even considered as the grace of God that we have been saved from those lower levels and have reached the human stage where we can consciously choose and create our destiny. Yes, we are lucky to reach the level of humans, and having got this opportunity as a self-conscious being, we should make every effort to evolve to the next level (to liberate our soul from the human level) and help others to do so.

The Mechanism and Process of Evolution

Each of the seven levels or layers called rays of the body or light complex of the soul corresponds to one of the densities and has a unique pattern and frequency of that density. For experiencing life on each density, the corresponding ray has to become the primary ray or the activated ray which then determines the entity's particular state of existence (the body and the environment) and subsequent experiences. The red ray has the lowest frequency and highest wavelength whereas the violet ray has the highest frequency and lowest wavelength and the manifestation varies accordingly with each density. As per the Planck theory, a photon's energy is proportionate to its frequency and hence among the rays, the red ray has the least energy and each ascending ray has increasing energy with the violet ray having maximum energy.

At the soul level, the purpose of each density experience is to perfect the ray that is in primary activation (core vibration) and balance the lower rays, if necessary, and attempt to open and work on the upper rays. As and when the core vibration matches the higher ray vibration, the soul is ready for graduation to that density. Each of these seven rays corresponds to the seven colors (as given in the table below) that we get to see when white light is refracted through a prism. Rays contain the blueprint of a soul's entire life cycle.

Density #	State of awareness or consciousness	Primary Ray	Physical Manifestation
1	Simple awareness	Red	Matter/Energy
2	Growing awareness	Orange	Plants and Animals
3	Self-awareness	Yellow	Humans
4	Love	Green	Angels - with a dense gel/plasma body
5	Wisdom	Blue	Higher Angels - with a lighter gel/plasma body
6	Balance – Balancing love and wisdom	Indigo	Higher Angels with an ethereal body
7	Oneness - Gateway to Timelessness	Violet	Highest Angels – with a pure light body

For learning purposes, we can imagine the body or light complex of a soul evolved to the level of humans as a pipe (as illustrated below) that runs along the spinal column, starting at the bottom of the spine and reaching up to the crown of the head. Here we can identify each Ray and its Chakra (the focal point of each ray) having proximity and close correspondence with certain parts of the physical body. The cosmic energy or *prana* enters the pipe at the lowest chakra which is in the red ray (south pole). It then moves upward through orange, yellow, green, blue, indigo, and returns to the cosmos through the violet ray (north pole). Each ray uses a portion of this *prana* for its maintenance and development.

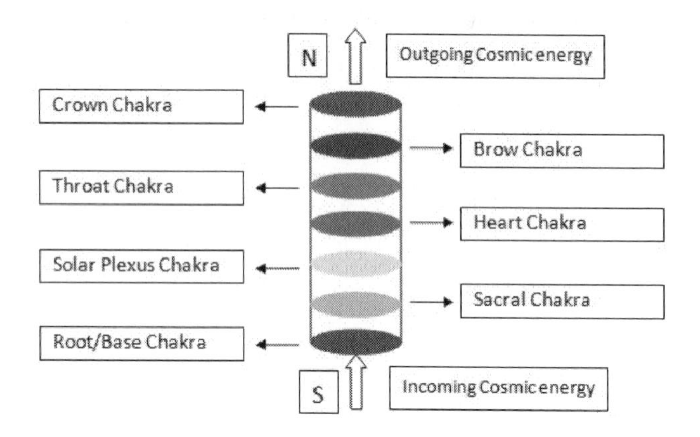

Concerning a human body, we can visualize that this energy enters the soul through the feet/ base of the spine and exits through the top of the head. As the entity evolves, its capacity to draw and use *prana* will increase. Therefore, in higher densities, the soul accumulates more *prana* and appears as more luminous.

The whole universe consists of souls progressing at various stages of evolution; appear as sleeping in the mineral kingdom, awake and moving in plant and animal kingdoms and self-conscious in the human kingdom and so on. The environment and experiences in higher densities are more refined as life becomes thought based without much need for physical activities. The higher the development of the soul, the higher is the level of its awareness and more the power of the mind.

At the initial stages, the mental capabilities are latent and mostly subconscious, and as the soul evolves, they come into operation in ascending degrees. Irrespective of the stage in existence, there is an essential unity underlying all souls, and each influence and is influenced by other souls and the Super soul continuously. As the incarnated soul evolves, the physical body it inhabits also undergoes certain transformation to adapt to the soul's vibration. The more harmonious the environment, the more rapid will be the evolution of the soul.

Reincarnation is the method available to the soul for evolving by experiencing a variety of life and learning the lessons within a certain density until its graduation or transition to the next level. The soul departs the body at the time of death and moves to the sub-dimension (time-space location) of the planet and after a period of review and acceptance of the just ended physical life it returns (which is almost instant in lower levels as there is hardly anything to review) to the proper dimension (space-time location), in the body of a baby, with specific targets/lessons to achieve/learn, and Karma to balance, based on its unique requirements. All our experiences are there to teach us something and restore harmony and balance in life and help us to evolve. Each experience teaches some particular lesson and certain experiences recur until the soul learns the intended lesson. Once the soul attains all the required learning and growth, then there would be no need for further experiences and hence no further incarnation in that density, unless it chooses to incarnate to help those who are left behind.

The graduation to higher density is determined only based on the vibratory rate of the soul and has nothing to do with the incarnated person's material achievements or status or religious affiliation in the world. If the soul did not achieve the requisite development during one density cycle, it will have no option but to repeat another cycle on the same planet if the conditions permit, else on another suitable planet. There is not much of a difference in the behavior of one soul who has reached the end of a particular density and another soul who is at the beginning of the next immediate density.

Apart from other factors, evolution is also influenced by the location of the planet within the solar system and the location of the solar system within the galactic system. The orbit of the sun is designed in such a way that it can periodically (in millions of years) gets closer to the galactic center. As the sun accelerates through the cosmic path carrying together all its planets, the closer it gets to the galactic center, the faster the evolution becomes, and the farther it gets from the galactic center, the slower the evolution becomes within the solar system. This can be compared to the seasons on our planet where we periodically get exposed to different climatic experiences.

The whole mechanism and process of evolution is about the soul seeking vibrational alignment with the Super soul in ascending degrees. Evolution leads to growth, development, and balancing of energy centers or energy bodies or rays and the mind complexes of the soul.

> ***Note**: We have a conscious time-space (metaphysical) experience when we enter the time-space dimension after physical death. Whereas, the sub-conscious time-space experience can occur to us at any time, and mostly during dreams.*

Stages of Evolution

First Density – Density of simple awareness – Matter/ Energy

The physical universe comes into being when souls manifest as photons and join together to make different forms of matter in its five elementary forms/ states such as solid, liquid, gas, plasma, and ether. In the case of a solar system, Sun is the first object to appear which then causes the creation of the planets. Scientists have discovered that Suns at their young age cause much chaos and help the elementary particles to coalesce and form planets and other celestial bodies. The first density is the density of simple awareness, in which the souls move out of the timeless state into physical manifestation. Sun, being a co-creator, is part of all densities or represents all densities of the major cycle and supports the souls to evolve (the planets and the individual beings physically and metaphysically).

When the physical universe is formed it is considered as the first stage of the evolution of souls and generally called the mineral kingdom (the inanimate matter/energy stage). Core density or vibration of the souls and the planets at this stage will be at first density. This stage takes millions or billions of years to evolve (estimated as about 2 billion years in the case of our planet). In this stage, the primary ray or the activated ray is Red and the souls that appear as matter do not show any sign of livingness outwardly. In this density, the mind and spirit complexes of the soul are mostly dormant or subconscious. First density souls are the material for creating physical bodies of the universe. The purpose of the first density is to build a foundation and an environment for further growth and development of the soul.

Nothing in the universe is as solid as it appears to be because they all consist of fundamental particles that are in the waveform. The seeming solidity of a thing is created when atoms stay together as a certain form for a particular period based on natural laws. The form of matter can be changed if its atoms/ subatomic particles are induced to rotate/ vibrate at different speeds. By the end of the first density stage, the earth became a solid sphere adorned with oceans, mountains, and atmosphere. One example of souls' evolution in first density is the transformation of carbon into diamond.

Second Density – Density of growing awareness – Plants & Animals

Towards the end stage of the first density period, the core vibration of the planet and some of the souls (after billions of years of its interactions with other souls, and the cosmic energy) move into Orange Ray bringing the mind complex into action and these souls get released from the matter stage. Over a period of time, the basic faculties such as organs of action and organs of senses come into operation in ascending order and these evolving souls manifest as the vegetable kingdom and animal kingdom by creating a variety of bodies. In the initial stages of second density, there were only unicellular organisms which later evolved into more complex biological forms over a period of millions of years (Microbes not only evolved to more complex beings but also created the fundamental ecosystem for all other second density and third density life on the planet). It is estimated that the earth remained a second density planet for about 4 billion years.

Second density is about the soul growing in awareness and striving towards self- awareness. In this density, the soul enters its growth phase and can freely move in space-time (during incarnation) as well as time-space (while discarnate). The behavior of entities is mostly subconscious or instinctual. During the second density experience and learning, some souls' spirit complex becomes active and they begin their work on the yellow ray. In these evolved second density souls there will be a certain level of awareness with basic mental activities, as can be seen in some pet animals, chimps, dolphins, etc. As the intellect available to these beings is limited to their low level of awareness, they have no capacity for rational and abstract thinking. However, except for rational and abstract thinking, many highly developed animals can handle most of the mental activities that an ordinary human being can perform.

The purpose of second density is to introduce the soul to basic life experiences where it has to deal with other selves, mostly for its survival and reproduction, so that it can evolve and move towards a self-conscious life.

Third Density– Density of self-awareness – Humans

Third density begins when the core vibration of the planet and some of the second density souls move to yellow ray. This is the density of self-awareness where souls experience life as human beings. To become self-aware, the soul needs to be capable of rational and abstract thinking. Hence, we can see that only those second density beings with a fairly developed mind can move to third density. We, humans, are not only aware of our environment but we are also aware that we are aware of it.

Those second density souls that have become capable of rational and abstract thinking graduate to the third density existence and continue their evolutionary journey using those advanced second-density bodies available then, which can reasonably support them (in our case it started with the ape body). When the second-density physical/ chemical body gets influenced by the third density vibrations of the indwelling soul and the third density planetary vibrations, it gradually transforms into higher configurations. Thus, the ape body gets transformed into that of a human body over a period of time. This corresponds with the Theory of Organic Evolution.

In the third density, the souls have become self-conscious for the first time in its evolutionary journey. Here the activated spirit complex and the developed mind intensify the evolution of the soul towards higher experiences of life. The individual begins third density with a feeling of separateness and lack of self-respect. Consequently, he or she behaves very selfishly without any regard for the rights of others, mostly guided by greed and fear. Those who are powerful try to dominate and subdue the weaker ones and exploit them. However, this selfish behavior gradually changes over many incarnations in a positively polarized soul as it spiritually progresses and becomes a selfless soul towards the end of its third density existence.

Before entering the human life, the soul had to evolve through millions of species and thus accumulated all those emotional energies (both positive and negative) of past lives, slightly modified by each subsequent life. This means that the immediate previous life before graduating to third density does have a tremendous impact on every human being. Therefore, the behavior of a human being, who is not evolved, can easily descend to the level of an animal, which is true in the case of negatively polarized individuals also, whereas, a positively polarized human being who is about to complete the third density can manifest great love and compassion like a positive angel.

The third density duration is very short compared to other densities. On earth, it is less than 100 thousand years (with one incarnation of about 100 years at present) to mature enough for graduating to the next density which is the fourth density. However, for each soul, the actual time spent may be different due to variation in opportunities available and used, as there is a great emphasis on the use of 'free will' because, human souls, having the intelligence to know what is right and wrong, can consciously evolve if they choose to do so. This density is also the density in which the souls make a choice between positive and negative paths for fourth density evolution. This choice or selection is a kind of subconscious one based on the pattern and frequency of the soul. The primary characteristic of a positively oriented being is 'selflessness' manifesting as love and concern for others whereas in a negatively orientated being it is 'utter selfishness' manifesting as domination and control.

Third density experience is basically for the development of the soul as a balanced unit of body, mind, and spirit complexes and then for choosing the polarity for further evolution. Though the core vibration of the soul will be in yellow ray, these souls have all the higher rays open or available for work and can work on them if they choose to do so, consciously or subconsciously. During this density, the primary focus of the soul will be on social relationships.

Fourth Density – Density of Love – Angels

The fourth density manifests when the core vibration of the planet and some third density souls moves to green ray, where the primary nature of awareness is love. When a soul has attained the requisite vibration of love (either love of others or love of self), the core vibration moves to the green ray (heart chakra), and it is ready to incarnate as an angel on a fourth density positive or negative planet. For those souls who have chosen the positive path, fourth density experiences are of love, understanding, compassion, harmony, and bliss. Life is predominantly community or group based as all are working with love and unity without any selfish goals. However, for the negatively polarized souls, the experiences will be the opposite of positive souls as their actions are based on 'self-love' rather than selfless love. The purpose of fourth density is the development and balancing of the green ray and working towards higher levels of wisdom.

This is the density where souls manifest as angels and begin the experience of higher awareness. This higher level of awareness is beyond our current comprehension, but we get a glance at it during vivid dreams. The activities are predominantly mental (thought-based) whereas for an evolved human being it is a mix of physical and mental activities. Here we can manifest thoughts instantly and move around at will. We can communicate with anyone anywhere as if we are present everywhere - beyond the limitations of space and time to a certain extent. It is an open density where nothing is hidden - each one can see/ understand the thoughts of others. The physical body for this density is made of different chemical elements (plasma) and hence is lighter, crystalline, and flexible compared to what we have now. This density has a length of few million years and the duration of one incarnation is a thousand times more than that of humans. This is the next immediate stage for us. This is the heaven many are looking forward to.

The core vibration of our planet has already moved into fourth density positive and we are in a transitory period now. In the book of Revelation in the Bible (which is a collection of visions subject to editing and different interpretations), there is a mention about 'new sky and new earth' which description matches the fourth density transformed earth as fourth-density life is experienced above the ground, on a thick layer of clouds enveloping the entire planet.

Fifth to Seventh Densities – Higher Angels

These are the higher densities that exist beyond our fourth density existence which are far away in our future to worry about now. However, for the sake of completeness, some details available about these densities are shared below:

Fifth density manifests when the core vibration of fourth density planet and some of the fourth density souls move to the blue ray. Blue ray is about wisdom and in this density experience, souls complete the lessons of wisdom by gaining knowledge of all available truths. This density spans millions of years, probably without the need for reincarnations as the soul can materialize and dematerialize the body at will. As the overall vibration of the soul has reached a higher level, it manifests in a more radiant and flexible plasma body than fourth density angels. In this density, a soul is able to operate with much more freedom and power compared to fourth density. While fourth density positive souls are naive because of their overly compassionate nature, the fifth density souls are considered wise.

Sixth density manifests when the core vibration of a fifth density planet and some fifth density souls moves to Indigo ray. The sixth density is the density of unity where love and wisdom are balanced. Souls manifest in a flexible ethereal body without the need for reincarnations. Also, the split of positive and negative paths that happened in the fourth density is unified during this density. This density lasts for a few million years. By the end of this density, the soul is removed from all traits of ignorance and becomes a conscious co-creator. Within our solar system, planet Venus has reached the sixth density vibrations as of now.

Seventh density manifests when the core vibration of a sixth density planet and some of the sixth density souls reaches Violet ray. This density is the highest or purest vibration of creation and the planet and souls will be in complete alignment with the vibration of the co-creator/Creator. The seventh density is considered as the gateway density, as towards the end of this density, souls lose the idea of separate existence and return to its original state before creation and merge with the Creator for a timeless existence which can be considered as the eighth density. This density may last for a few billion years. Here the manifestation is nothing but the light of the highest frequency. The planetary conditions will be like the Sun. Within the solar system, so far, no planet has evolved to seventh density. However, Sun, being the co-creator, has seventh density also.

Notes:

1)First density to third density life is gross body-based and lives on the ground of the planet whereas the fourth density life and above exist above the ground.

2)Each ascending density will have a physical body that is more flexible, crystalline, and translucent than the previous one.

****

Chapter IV: We, the Humans

The evolutionary stage of souls as humans. The constitution and various connections and interactions within and without. How the soul, specifically the mind, manages the body using the nervous and endocrine system. The purpose of our human existence and how we can consciously align with it.

Introduction

In the previous chapters, we had a general introduction about the soul and how it evolves through various densities of existence and finally merges with the Creator at the end of a major cycle. Through this chapter, we are going to specifically examine the souls who after billions of years of evolution on planet earth, have reached the level of humans.

This section explains our constitution and life as a human being and how our inner and external environments influence each other every moment of our life. It also explains the purpose of our incarnation and how we can consciously evolve to higher levels.

Our constitution/structure

As discussed in the earlier chapters, we are a soul presently experiencing life on earth with the help of a physical/chemical human body. Let us examine each part of our constitution, starting with the gross/physical level:

Physical/ Chemical body – our temporary part

The physical body, with all its complex organs and systems, consisting of the skeletal system, muscular system, nervous system, endocrine system, respiratory system, circulatory system, digestive system, reproductive system, etc., is the instrument that helps us to experience our human life on this planet. This body is made up of trillions of cells created from the molecules that are formed when certain types of atoms join together. These cells, though having a certain level of livingness, intelligence, and independence (as they are also souls at the initial stage of their second density existence), do different kinds of jobs individually and in groups, as assigned to them by the controlling intelligence of the indwelling human soul and serve the purpose of providing a body/ vehicle to a higher soul.

At the fundamental level, we can consider the body as an organization of cells working together for a common purpose. All the cells in our body from head to toe are replaced periodically. It is estimated that more than 50 billion cells die, and a similar number are born (reproduced through cell division) daily within a healthy adult human body. Through this process, on an average, once every year the whole body renews itself. The health and survival of the body are dependent on this constant renewal. If this renewal is delayed or stopped then disease and death of the body would follow. These cells are assisted by trillions of microbes (also second-density beings) in carrying out various activities within the body for its smooth functioning and maintenance. Each cell is constantly on the alert to build, repair, protect, and work in every possible way to keep the body in good working condition.

To understand the functioning of the physical body as a unit, we need to first understand the nervous system, which is the communication and control system within the body that handles the stimulus-response mechanism under the overall control and supervision of the mind. The nervous system is that part of the body through which our soul connects and communicates with the body. The nervous system is broadly divided into two parts - the central nervous system (CNS) and the peripheral nervous system (PNS). The central nervous system is the command center that includes the brain and the spinal cord. The peripheral nervous system is made up of two parts - the somatic nervous system and the autonomic nervous system. The somatic nervous system controls nerves of the sense organs whereas the autonomic nervous system controls the nerves of the inner organs of the body. The neuron or nerve cell is the fundamental unit of the nervous system. The neuron is a specialized conductor cell that receives and transmits electrochemical nerve impulses. There are several types of neurons. Two major types are sensory neurons, that take messages to the nerve centers, and motor neurons, that carry messages from the nerve centers.

At the primary level, the function of the nervous system is to send and receive signals from one cell to another cell or from one part of the body to others. There are multiple ways a cell can communicate with each other. One such way is by releasing chemicals called hormones into the internal circulation/ blood so that they can reach every part of the body. This is accomplished by making certain glands to produce and release chemical substances (hormones) that are not related to their organic processes. Another way is by neural signaling (radiation) where cells directly communicate with other cells. At a secondary level, the function of the nervous system is to manage/control the body by working as a communication system/network between the body and soul. It accomplishes this work by collecting, organizing, and exchanging information between the body and soul.

Though the physical body is not our permanent residence or vehicle, we should understand and take care of it because without it we cannot experience the earthly life and work out our Karma. We are an individualized unit of God; the physical body we inhabit is considered as the temple of God and we need to take proper care of it so that we could live happily and productively for the entire intended period of an incarnation. A driver needs an automobile in working condition to reach a certain destination. However, an automobile in perfect condition can guarantee a smoother, faster, and enjoyable ride too. Similarly, a soul needs a working body for basic survival on this planet, and a healthy and disciplined body enables it to experience abundant life.

Notes:

1) Most of the traditional healing methods attempt to heal the physical body only.

2) It is strange that many values and take care of temporary material things like job, car, house etc. better than their bodies.

Soul – our permanent part

The soul is a subtle energy web comprising of three distinctive and hierarchical levels or complexes viz. light body (rays), mind and spirit, as explained below:

1) **Light body** is the metaphysical body of the soul, consisting of seven layers called rays, each corresponding to one of the seven densities or vibrations or frequencies of creation. These seven rays, its chakras, and their functions are explained as a separate section later in this chapter as it needs to be dealt with in detail.

   These rays are light which is in constant vibration and trying to keep its optimum balance while incorporating the impressions or messages it is continuously receiving from/through the mind. Its vibration mirrors or reflects the activity of the mind and each ray has its manner of perception and represents specific feelings according to the nature and state of the mind it has been linked to. These rays seat the principle/energy of five sense organs (vision, hearing, smell, touch, and taste), five organs of actions (mouth, feet, hands, the organ of excretion, and organ of procreation), five objects of senses (taste, touch, sight, sound, and smell), mind and spirit.

   Each ray is associated with one of the seven nerve centers (nerve plexus) that are linked to certain endocrine gland(s) which manages certain organs/parts and/or functions of the physical body. Each ray interacts with the physical body through the nervous system. All our physical experiences are caused by and experienced at the level of one of these rays (with the help of mind) which then gets transmitted to the physical body and vice versa.

> ***Note:*** *Some energy healing techniques attempt to bring changes to this part of the soul without giving any regard to the mind which controls it. Hence lasting results are not forthcoming and sometimes create imbalances.*

2) **Mind** is that energy complex of the soul which connects the light body complex with the spirit complex and makes the soul work as a functional unit. It is the tool of communication and perception with which we can communicate, feel, think, remember, imagine, dream, seek, learn, understand, and also experience intuition and awareness. Mind has two levels called conscious and subconscious (this division appears only when the soul incarnates in a physical body), each with multiple sublevels connected to each ray of the light body and the spirit. The conscious mind handles current information and helps us to think rationally or logically and focus, whereas the subconscious mind handles past information (experience and knowledge accumulated over the entire period of the

existence of the soul) and guides us based on that information. While in a physical body, all the voluntary activities are handled by the conscious mind, and the involuntary activities are handled by the subconscious mind.

While within the physical body, human perception is a complex process involving sensory inputs collected through the sense organs and organized by the brain/ nervous system which are then picked up by the rays and identified, interpreted, and reflected by the mind. Information is not stored in the brain, but the rays and the mind are our tool for storage, retrieval, perception, and communication of information. To a certain extent, we can compare the nervous system to the hardware and the mind to the software of a computer system. Thoughts and feelings are those subtle energy waves or vibrations that emerge from or are directed by the mind.

The extrasensory perception (ESP) or sixth sense is the ability of the mind to know directly without the use of the sense organs. This direct knowing is also called intuition which can manifest as a spontaneous feeling or a thought in the mind. It is a spontaneous knowing of something that we never consciously thought of. This is the way we know many things from the subconscious mind, spirit as well as other minds (individual or collective). Intuition naturally guides us towards our goals/purpose.

Mind is conditioned by the information it handles and the condition of the connected rays, other minds, and the overall tuning it has attained with the spirit. Further, the information is handled by the mind according to its conditioning, which manifests as certain biases. Therefore, our perception or awareness is limited to our soul or spiritual development and balance and accordingly determines our reality. If we compare the mind to water, then a clear mind is like clear water and a conditioned mind is like muddied water. When we look at muddied water, we see nothing but muddiness. As the water becomes clear we start seeing the water itself and its environment also. When it becomes very clear then we can see everything clearly including our image and other images that reflect in it. In other words, a balanced and calm mind has more clarity which helps us to effectively seek the truth and find it fast.

The disharmonious or imbalanced state of mind is primarily due to the less developed and disharmonious or imbalanced condition of its immediate environment. An imbalanced environment (rays, physical body, and nature [including other minds] around us) makes the mind more and more imbalanced (agitated and confused or clouded) and over some time the individual even loses the capacity of right judgment/ perception and understanding. If such an imbalance or disharmony is allowed to continue, then there could be all kinds of problems including diseases in the physical body and mind. It also negatively affects the spiritual progress of the individual.

Mind is the center of every activity of our life and the mind waves can be directed, focused, or changed with each conscious or unconscious thought. While experiencing human life in a physical body, majority of the activities of our mind happen subconsciously. Negative thoughts lead to negative conditions and positive thoughts lead to positive conditions. When we know our mind better and use it wisely and take care of the environment, then we will be able to make meaningful progress in our life.

Notes:

1) Holistic systems like yoga and spiritual development programs attempt to calm the mind by clearing the conscious mind from random thoughts by making the energy movement within the soul harmonious and not feeding it with new thoughts. Since it is more practical and sensible to heal the soul rather than healing our external environment, which is mostly beyond our control, spiritual development programs have an advantage over other methods and systems for our overall wellbeing.

2) Healthy environment supports a soul to lead a healthy life. Our physical body, family, friends, and other close contacts, the culture, and climate of the place of residence, etc., play an important role in our evolution or spiritual development.

3) **Spirit** consists of the purest or least distorted portion of the intelligent energy of the Creator and is considered as the co-creator within the individual soul. It is also known as the inner light and it resides within the Indigo Ray. This energy complex acts as a channel between the individual soul and the Super soul (Creator) and keeps the individual soul focused on the ultimate purpose of life. When the body and mind complexes do specific tasks for the existence and experience of the soul as an individual being, the spirit is the experiencer who remains unaffected by its surroundings and simply expresses the nature of the creator.

   Depending on the level of development of the light body and mind complex, the spirit complex can effectively function as a channel or link between the individual soul and the Super soul and reflect or partake more of the nature of the Super soul. Through the spirit complex, the oneness or the unity of 'souls' is maintained - a Creator that has become many individual units and yet remains as one. When we realize the true nature of our being and the oneness with the Creator then we know we are truly the 'children of God' and there is nothing that can separate us from God.

> ***Note:*** *We do not have to attempt to heal the spirit as it will automatically find its balance and express itself to the extent that our mind and body complexes are developed/balanced. As we evolve and reach higher levels of awareness, to that extent, we can partake of the nature and power of the Creator. When we talk about 'spiritual healing' what we mean is the removal of hurdles by healing/ tuning the light body and mind complexes.*

We are the soul or photon - the fundamental particle - that after billions of years of evolution, has developed both in quantum and quality, and has reached the stage of a human being. Every soul that has reached the level of third density has the capacity to discover its true nature and oneness with the Creator in an ascending order. This self-discovery is commonly known as 'self-knowledge' or 'self-awakening' or 'self-realization'. When we understand our 'self' we can understand other 'selves' (everything else) too because All is One.

As above, so below; just like the universe is multidimensional, the individual is also a multidimensional being. Each soul is a holographic version of the universe. Each soul contains within it all the densities and sub-densities and can exist in both space-time (physical level) and time-space (metaphysical level) using appropriate bodies. It is a group of energy complexes (electromagnetic waves) that interpenetrate, interact with, and influence each other. There is no absolute border between these complexes as they shade into each other, and blend like the colors of a spectrum. There is an inherent harmony, agreement, and correspondence between all these complexes, and one cannot continue without the other.

Though there is a certain level of independent actions within each of these complexes, body, mind, and spirit are hierarchical in nature and spirit is the true self who is in command of the soul. Therefore, the energy flow is from the spirit complex to the mind complex and mind complex to the light complex and vice versa. Further, each of these complexes is connected to and influenced by its corresponding level in the universal scheme and also ranks hierarchically, and the higher levels incorporate and support the lower levels. Therefore, everything in the universe appears as hierarchical, connected, and is influenced at every level by each other.

Our life experiences are consequent to the interactions of all three complexes and not just caused by any individual one. The light body (rays) is the permanent body of the soul within which the mind and spirit reside and is our gateway to the physical world. Spirit is the master that gives direction and also acts as the link with the Super soul and other souls. Mind is the center of activity within the soul that connects the body complex with the spirit complex and helps the soul to work as a unit.

Notes:

1) Souls normally take full charge of a new physical body closer to the delivery time and in some rare cases a few hours after the birth. Death occurs when the soul leaves the body.

2)The words 'light' and 'energy' are used interchangeably because the light is the basic form of energy.

3) Aura is the electromagnetic field that emanates from the soul and surrounds it. As the soul evolves, the quantum and quality of the aura also increase.

Rays and *Chakras*

Rays are the energy complexes or electromagnetic or light fields that form the subtle or metaphysical body of the soul. As per Hindu philosophy, Rays are known as *Koshas* in Sanskrit (translated as Sheaths in English) and considered as a covering or garment of the *Atman* (spirit) and exist in multiple layers. Chakras are the focal points or focus areas of the rays that act as the entry/ exit points (valves) regulating the flow and interaction of *energy, including Prana. Chakra* is a Sanskrit word meaning 'wheel', because the energy is supposed to be rotating like a wheel or vortex. There are seven Rays and accordingly seven major chakras in a soul. Let us examine each layer/ray of the subtle body of the soul and understand its nature, functions, and connections with reference to a human body.

1) Red Ray

Red Ray is the foundation ray of the soul. It is the first density manifestation within the soul. It is the elemental body of the soul that consists of the blueprint and substance for the creation of all forms of bodies. This ray processes or is concerned with the catalysts of the survival of the body.

While in human incarnation, the basic concerns of this Ray are meeting the physiological requirements such as air, water, food, shelter, and safety of the physical/chemical body (specific survival requirements may vary from density to density). DNA molecule within each cell containing the genetic instructions is a physical manifestation of Red Ray activity ensuring the survival of organisms.

The name of the chakra is the Root chakra and its location corresponding to the physical body is the base of the spine. Organs associated with this chakra are the kidney, large intestine, and bladder. The major nerve center associated with this *chakra* is the Lumbar plexus and the linked gland is the Adrenal which is responsible for blood pressure and stress management, especially the 'fight or flight' response. The major hormones secreted by this gland are Cortisol and Adrenaline.

Root *chakra/ray* can be blocked or imbalanced if there is a fear of survival when the individual is not in a position to adequately meet his basic physiological requirements or is going through life-threatening situations. Imbalances/blockages appear as fear, anger, aggression, indifference, lack of self-respect/ acceptance, selfishness, addictions, suicidal tendencies, helplessness, escapism, lack of motivation, depression, and a weak body. Any imbalance/blockage in this *chakra/ray* leads to blockages in all upper *chakras/rays*. Individuals with a clear/balanced root chakra/ red ray appear as healthy, cool, flexible, brave, stable, and confident.

2) Orange Ray

Orange Ray represents the second-density within the soul. This ray is concerned with reproduction and related matters. Once the basic survival needs (red ray needs) are relatively satisfied, the sexual and personal relationship/emotional needs like the need for a partner or friends comes into play.

The name of the *chakra* is Sacral *chakra* and it is located in the pelvic area. This chakra is associated with the reproductive organs. The major nerve center associated with this *chakra* is Sacral Plexus and the linked glands are Gonads (*ovaries in women and testes in men)*, which produce the sex hormones and manage the reproductive system.

This *chakra/ray* can be imbalanced/blocked when the individual is unable to maintain a healthy sexual relationship, or the sexual/reproductive needs are unmet. Such individuals have an insatiable craving for sex and appear to exist primarily for having sex. Individuals with a clear/balanced Sacral chakra/Orange ray have a healthy sexual life and are able to exercise self-control.

> ***Note:*** *Like survival needs, sex is also a basic need of the individual (as it pertains to reproduction), which is difficult to suppress, especially when the entity is young - when it has the maximum capacity to reproduce. As the entity matures and evolves, to that extent this need reduces as the needs of higher rays take precedence.*

3) Yellow Ray

Yellow Ray represents the third density within the soul. This is the ray that is in core vibration for us now. This ray is primarily concerned with the individuality and social needs of the individual. Activation of this ray enables the soul to experience a self-conscious life.

As the soul has become self-conscious it is not only aware of the environment but also aware of itself and its roles and responsibilities as a part of the society/universe. Once the red and orange ray requirements are relatively fulfilled, the individual feels the need for acceptance and belongingness in the society. At this stage, the focus will be on ego and relationships like family, friends, neighbors, religion, politics, and other relations in society.

The popular name of the *chakra* is Navel *chakra* and its location corresponding to the physical body is at the upper half of the belly. Associated organs are the stomach, spleen, pancreas, liver, gallbladder, small intestine. The major nerve center associated with this *chakra* is the Celiac plexus (also called Solar plexus) and the linked gland is Pancreas, which releases a hormone called insulin that helps cells to take in glucose as its fuel.

A perceived threat to the individuality of the person and social relationship issues can create blockages/imbalances in this *chakra*. Blockages/imbalances appear as anger, hatred, egoism, resentment, inflexibility, inability to accept the rights of others, and lack of respect for the social norms and the law of the land.

Individuals with a clear/balanced yellow ray can accept the role of others in their lives and successfully engage with them individually and in groups and appear as flexible, motivated, practical, and content. They have self-control and take responsibility for their life. Positive individuals are humble, friendly, kind, loving, and generous whereas negative individuals are selfish, devious, overly materialistic, manipulative, and dominating.

4) **Green Ray**

Green Ray represents the fourth density within the soul. Green ray has access to deeper levels of mind and spirit and can express the fundamental nature and power of spirit, which is love.

The name of the *chakra* is the Heart chakra and its location corresponding to the physical body is the chest. The major nerve center associated with this *chakra* is the Brachial plexus and the linked gland is Thymus, which produces thymus hormone and manages the immune system. This chakra is associated with the heart.

Only those with a fairly balanced lower rays (red, orange, and yellow) can begin their conscious work on this ray and look forward to graduation to fourth density.

Individuals of positive polarity with an open-heart *chakra* are able to love others without expecting anything in return (unconditional or universal love). They appear as transparent, forgiving, compassionate, and trustworthy. They can see the creator in self and other-selves and treat others as part of themselves. Whereas, individuals with negative polarity love themselves to the exclusion of other selves, consider themselves as special, are manipulative, cruel, and cunning. For them, others exist for their comfort and pleasure. They are obsessed with power and are ready to go to any extent to achieve their goals.

Blockages/imbalances appear as jealousy, alienation and loneliness.

Notes:

1) *We always want to be loved because that is our true nature and that is when we feel good.*

2) *Though we all love ourselves and many of our actions are based on self-love, most of us have compassion and concern for others and hence are not willing to harm others knowingly, which makes us a positive being.*

5) Blue Ray

Blue Ray is the manifestation of fifth density within the soul. This ray is primarily concerned with knowledge (wisdom) and its expression (communication).

The name of the *chakra* is Throat *chakra* and its location corresponding to the physical body is the lower half of the neck (throat level). The major nerve center associated with this *chakra* is the Cervical plexus and the linked gland is Thyroid, which produces thyroxin hormone and manages certain metabolic functions, growth and development, body temperature, respiratory system, and speech organs.

This ray houses all the fundamental truths that the soul has learned throughout its evolutionary journey and guides the soul to further truths. Those who have a fairly balanced red, orange, and yellow rays can work on this *chakra.*

Individuals with a blocked throat *chakra* have difficulty in learning fundamental truths and appear as ignorant, superstitious, fanatical, and frustrated. Those with a clear throat *chakra* are seeking, understanding, intelligent, independent, and dare to express themselves without bothering for the consequences. Positive individuals appear as honest and straight forward. They inspire and teach others without expecting anything in return. Negative individuals are dishonest and crooked.

6) Indigo Ray

Indigo ray is the sixth-density representation within the soul. This ray houses the spirit complex within the soul.

The name of the chakra is Brow Chakra (also known as the 'third eye') and its location corresponding to the physical body is the forehead, between the brows. The nerve center associated with this chakra is Cerebellum and the linked gland is the Pineal, which produces melatonin hormone. This gland manages top-level communication between the body and soul. It controls the body's clock and circadian rhythms.

The indigo-ray is available for conscious work only to those evolved entities who have understood and accepted their true nature and are able to live with that realization. Blockages appear as limited awareness/confusion and consequent lack of interest in self-development/ self-realization/ spiritual matters.

Those with an open brow chakra experience higher levels of awareness, strong intuition, peace, and bliss. They consider themselves and others as co-creators. They are balanced (wise and loving) and able to live in harmony with the physical and metaphysical worlds.

7) Violet Ray

Violet Ray is the manifestation of the seventh density within the soul. Its energy represents the totality or beingness of the soul, which is being affected by all kinds of activities from each ray of the soul and its various relationships.

The name of the *chakra* is Crown *Chakra* and its location corresponding to the physical body is at the upper head towards the backside. The nerve center associated with this *chakra* is the Cerebral cortex and the linked glands are Hypothalamus and Pituitary. The hypothalamus controls the autonomic system that manages hunger, thirst, sleep, and sexuality. Hypothalamus also regulates body temperature, blood pressure, emotions, and secretion of hormones. It controls the Pituitary gland also. We feel happy and elated when the hypothalamus releases a hormone called dopamine. The pituitary is the master gland that controls other endocrine glands in the body and particularly manages sexual development, promotes bone and muscle growth, and responds to stress.

This ray always remains open but is not available for any work. It just reflects the overall condition/ vibration of the soul and we feel there either satisfied/fulfilled or dissatisfied/depressed. It automatically changes based on the changes happening to other rays.

For easy reference, all the Rays, their respective *chakras*, and their function areas are summarized below:

Ray #	Ray Name	Chakra Name	Functional Area
1	Red Ray	Base or Root *Chakra*	Survival
2	Orange Ray	Sacral *Chakra*	Reproduction
3	Yellow Ray	Solar Plexus *Chakra*	Relationships
4	Green Ray	Heart *Chakra*	Love
5	Blue Ray	Throat *Chakra*	Wisdom/Communication
6	Indigo Ray	Brow *Chakra*	Spiritual/Self-realization
7	Violet Ray	Crown *Chakra*	Beingness

The following table summarizes how the Rays manage the physical body through the nervous system:

Ray Name	Nerve Centre	Associated major gland	Important functions of the gland*
Red Ray	Root Plexus	Adrenal	Managing blood pressure and stress
Orange Ray	Sacral Plexus	Gonads	Managing reproduction
Yellow Ray	Celiac or Solar Plexus	Pancreas	Converting food to fuel
Green Ray	Cardiac or Heart Plexus	Thymus	Managing immunity
Blue Ray	Cervical Plexus	Thyroid	Managing metabolic functions, growth and development, body temperature, respiratory system, and speech organs.
Indigo Ray	Cerebellum	Pineal	Manages top-level communication with the soul. Controls the circadian rhythm, and reproduction.
Violet Ray	Cerebral Cortex	Hypothalamus	Controls the autonomic system and manages the Pituitary gland.
		Pituitary	The master gland that controls other endocrine glands in the body.

*This area requires further research

There is a rhythm or balance in the pattern and frequency of every ray of an evolved soul. In lower densities including third, due to various reasons, imbalances/blockages, and consequent reduction in the flow and use of *prana* in certain rays or all the rays is common. These imbalances/blockages have a corresponding effect on the physical body while in the incarnation. Some imbalances/blockages may last for few seconds only whereas some last for an entire incarnation or many incarnations. The severity of the imbalance/blockage and its duration determines how much loss or damage it can cause to the entity spiritually, mentally, and physically. As a general rule, negative feelings cause imbalances whereas positive feelings can balance them.

Rays work together as a coordinated system, each one having important and significant connections with the metaphysical and physical world corresponding to its density. Red ray is the foundation ray for our physical experiences whereas Indigo ray is the foundation ray for our spiritual or metaphysical experiences. Violet ray manifests the aggregate condition (manifests as a certain vibrational frequency) of all the other rays or the totality of the soul, which is constantly being changed by each thought, condition of the physical body, and by the numerous interactions between the individual and the universe. Other rays specialize in a certain area of our lives and together make it possible for the soul to exist as a balanced individual both in metaphysical and physical worlds.

As the soul functions as a single unit, we cannot consider the ray/body connections as an exclusive affair. However, understanding these specific connections will help us to identify specific blockages/imbalances/catalysts and handle them appropriately. Any attempt to forcibly increase the vibration of higher rays while the lower ones are imbalanced could cause imbalance to the soul and manifest as physical or mental diseases.

> ***Note:*** *Other than the seven major nerve centers there are several small nerve centers in our body, some known and some yet to be known. More research is needed in this area.*

Cosmic Energy

Cosmic Energy or *Prana* or *Chi* or Reiki is the fundamental energy of the universe. It appears in the universe as the feminine or *yin* aspect of the Creator. It is the photon field, in which every being lives, moves, and has its existence. Just like air is essential for the physical body to survive, *prana* is essential for the soul. Like air, it does not differentiate as to what use it is being put to. It can be used constructively or destructively depending on whom and for what purpose it is being used.

The soul inhales *prana* into itself from Nature through the south pole (red ray) and from there it moves upward. Each ray uses a portion of this *prana* for its maintenance and development. Spirit, the masculine nature or *yang* aspect of the creator at the universal level which resides in the Indigo ray, connects with the incoming *prana* on any one of the higher *chakras* (starting from heart *chakra*) and after the union, the remaining *prana* is exhaled through the north pole (violet ray). This is a continuous process like inhalation and exhalation by our lungs. (The union of *prana* and spirit becomes part of our life only after the soul has become self-conscious, and the resultant energy /transformation is fundamental to our spiritual development and work in consciousness).

Every moment there is an entanglement of *prana* and spirit happening on one of the upper rays. Only the location of this union varies depending on the ray which is in action and/or calls for the union at that moment. As the individual spiritually evolves, the meeting point moves up, and correspondingly the awareness, feeling/ experience of the individual will be of a higher order. *Kundalini* rising or *Kundalini* awakening/ experience as per the Hindu philosophy is a reference to this union and meeting point.

The use of *prana* by each ray can be compared to the use of electricity by an electrical appliance such as a bulb or a motor. As the electricity passes through the device it gets transformed into light, heat, kinetic and other kinds of energies. Similarly, within the soul, *prana* gets transformed into vital energy (life force), mental energy, magnetism, etc., depending on which ray is processing it. The requirements of each ray and the tuning/ alignment, the rays, and the connected mind have achieved with the spirit, determine the nature and amount of cosmic energy used by each ray, and the consequent experiences of the soul

Each interaction with *prana* not only adds more energy into the rays but also transforms the pattern and vibrational frequency of the rays and their linked mind energies and leads to certain feelings and emotional experiences. Further, this emotional state or vibration sets in motion a chain of actions, as this will be picked up by the connected nerve centers of the physical body, processed by the central nervous system, and passed on to the respective glands which will then produce and release certain hormones to alter the state and certain functions of the body. Through this mechanism, every change within our soul makes a corresponding impact on the physical body and also on the surrounding environment. Spiritually evolved entities can process large quantities of *prana* into positive mental energy and use them not only for themselves but also for inspiring and energizing others. Many energy healers know how to consciously use *prana* for various healing purposes.

Since the brain is the central part of the nervous system, it gets more affected than other parts of the body and changes every moment to reflect our soul's state. Neuroscience has proven that we can change our brains just by thinking differently. This is because when we change our thoughts or focus, first our metaphysical environment incorporates this change, and then the physical environment, starting with the brain, also changes to reflect this change. Therefore, every time a person experiences or learns something, that understanding brings a corresponding change to the inner and external environment of the individual. This is the mechanism of truths setting us free. With the proper use of the mind, humans can understand their true nature and make the right choices to live an abundant life and make meaningful progress in their evolution.

Life as a human being

As human beings, we are presently experiencing the 3rd dimension or density of life on planet earth. Though third-density experience is a much shorter period compared to other densities, it is a crucial density as this is where we make our choice for a positive (selfless) or negative (selfish) path of fourth-density graduation. This choice is the reason why we are surrounded by an ocean of experiences which is often confusing as we are presented with multiple choices each moment of our life. Only an awakened person is aware of the intensity and purpose of this environment whereas the majority have no clue about it and just go about their daily routine of a mechanical existence.

Upon our physical death, our soul departs the body carrying with it all the experiences and knowledge acquired during an incarnation. This soul will be automatically drawn to its place of rest/ healing in the sub-density realm that matches or resonates with its vibrational frequency without any kind of external intervention like a magnet attracts iron filings. After the due period of resting/healing, we reincarnate in the body of a baby, in total forgetfulness (because of the veiled subconscious mind) of our earlier existence. A soul may incarnate as a biological male or female as per its specific experiential and learning requirements. The cycle of birth and death in third density continues until the soul has attained the requisite vibrational frequency for graduation or transition to the next density, which is the fourth density.

At the time of physical death, we just lose the physical/ chemical body and come out of our forgetful nature and are able to see our past and present and determine the future incarnation with or without the help from spiritual guides depending on how evolved we are. Each incarnation is designed in such a way that there are enough opportunities for correcting the imbalances of the soul (specific rays). Consciously knowing the past lives or the future while inhabiting a physical body is normally not probable/ available in third density. However, this is partially possible during dreams, out of body experiences, and hypnosis.

While in a physical body, we perceive our world mostly based on the inputs of the sense organs processed by the brain/ nervous system and interpreted by the mind. Messages from the environment are interpreted and reflected by the mind according to the overall condition of the soul under the influence of the ray in which the individual is primarily operating/ focusing his attention at that moment. This means that an individual's perception is limited by the level of development his soul has achieved, as he can see only what he is able or ready to see.

Mind responds instinctively to the stimuli that it receives from the environment consequent to which we experience or have spontaneous thoughts, feelings, and emotions. This instinctive reaction sometimes forces us to do what we do, even though we do not consciously want to do it, and sometimes does not allow us to do what we want to do. The instinctive reaction is based on the conditioning of the mind with which it subconsciously sees and reacts to everything. All our thoughts get colored with that conditioning just like what happens when we look through colored glass. Sometimes the individual tries to resist or suppress these uninvited thoughts or feelings or emotions for a certain period but they return with more power making the person frustrated and restless. Such disharmony, if allowed to continue, is the reason behind all our sufferings.

Two individuals may encounter similar kinds of events or sensory inputs, but the interpretation and response could be totally different for each person. An evolved person can see more clearly and thus respond in a balanced manner, because he has the support of a developed soul for better interpretation and understanding, whereas an ordinary person could just react to that event. Therefore, an evolved person is fairly insulated from the negative instinctive reactions of the mind. This is one of the reasons why we need to be forgiving and compassionate to our fellow beings, as many times they do not know what they are doing and why they are doing it, because of their lower awareness and conditioned mind.

Focused or sustained thoughts create feelings and emotions and when a soul is vibrating with a feeling/emotion that is negative/ destructive, such as anger or fear, it can also be interfered with and intensified by negative thoughts from other souls, both incarnated and discarnate (who have access to the energy system of the earth). In most cases, this targeted/victim soul is unaware of the intrusive thoughts as they are so mixed with his thoughts that he considers them as his own. Forcible manipulation by negative souls (only negative souls [also known as 'evil spirits'] breach the free will of another soul) needs particular mention here as they spontaneously communicate to the person telepathically and manipulate the thoughts/feelings in an attempt to make him or her commit certain evil deeds. Such negative acts satisfy not only the doer's feelings but also the feelings of the trespasser. This act of forceful communication and control by negative souls is generally known as 'spirit possession' or 'spirit attachment'.

During a spirit possession, not only the individual's thoughts are being manipulated, but his feelings/emotions are also being intensified by the collective power of negative thoughts. Thus, the possessed person finds it impossible to resist what he is forced to do and become 'insane' for a certain period of time. Compulsive behavior disorder, insanity, diseases, accidents, suicides, etc., could also happen due to spirit possession. Though the person has become a puppet, controlled, and manipulated by others from unseen levels, he is still responsible for the deeds committed through him and suffers accordingly.

If a negative state of mind continues for some time, it will eventually manifest as a mental and/or physical illness. The disease is the manifestation of the imbalance within the soul and gives us a clue as to what is going on within. While it appears that there are external causes of the disease, these external causes are there because of the inner causes only. As the person changes inwardly, the feelings/emotions that caused a certain disease get released, and the energy pattern and vibrational frequency return to its natural state, which is of balance and health. Our mind has the power to make our body both ill and well as our state of mind determines the state of our body functions, specifically the cells. If we are filled with love, then the cells love and support each other, and we experience health and if we hate others then our cells resonate with that nature and our body becomes diseased. This is the reason an evolved person is not easily diseased, and even if diseased, he can heal himself. Whereas, an ordinary person who struggles with certain instincts/ feelings/emotions will easily become diseased.

In certain cases, the physical body may be repaired or cured of a particular diseased condition with the help of medical science, but unless the mind is healed or balanced from the negative feelings or negative thoughts, the diseased condition will recur or manifest in some other way. This means, even though, use of modern medicine, especially surgical procedures, is necessary on certain occasions and has its important role in healing a physically diseased condition, it does not address or eliminate the root cause of the disease. Therefore, more than external intervention, it is necessary to balance the energy of the soul by replacing those negative thoughts/feelings that created the illness with positive thoughts/feelings. Most, if not all, of the causes of diseases, will naturally go away with inner change supported by natural living.

What we believe is true to us and becomes a reality in our lives and what we do not believe does not apply to us or gets released from our lives. This is because the subconscious mind is the creative mind that works constantly based on our beliefs to create or manifest reality. It is like a computer system that allows only those applications that are 'installed' on it. What is not installed is not available or recognized on that computer. One may wonder why many cannot recognize higher truths and enjoy a blissful life and instead live a miserable life because of ignorance. The reason is that what is true to one may not be true to another and what is true to a person is what matches or is in alignment with the level and frequency of that person's soul. Certain types of beliefs act as a filter or blockage to the subconscious mind and do not allow those ideas or concepts that do not resonate with the existing beliefs to enter it. These beliefs could be based on truths or lies or be imposed by the society (from tradition, religion, or law of the land that considers certain things as acceptable and legal and other things as not acceptable and illegal).

An ordinary person is just a believer – a believer of either lies, half-truths, or truths whereas an evolved person not only believes but knows them as part of his true nature and thus can relate to them. A wrong belief makes a fanatic out of a simple believer – both religious and non-religious - and a right belief can make a person, who is open and flexible, spiritual. Whereas fanatics believe something to the exclusion of everything else, a spiritual person is willing to hear the other side of the story, though he or she believes and cares for those things that are more meaningful and valuable and which can make life better, both here and hereafter.

As the person evolves, higher rays (green ray onwards) exert more influence on the soul, and the lower rays' need for energy diminishes as they are balanced and concerned with only essential and useful activities. Positive individuals who have balanced lower rays and have begun working on higher rays generally manifest compassion and humanity. They are concerned about the welfare of other-selves at a universal level without any separation/discrimination, whereas those who operate in lower rays exhibit animalistic behaviors and do things just for themselves and sometimes for those who are closely associated with them.

If an individual is struggling with survival issues, then he has an imbalanced red ray, and if he is struggling with his personal life, then his orange ray is imbalanced and if he is struggling with his social life, then his yellow ray is imbalanced. Those who are struggling with these rays are a slave to their lower nature/ lower mind and can hardly do any work on the higher rays. When most of humanity is struggling with survival issues, there is much difficulty in attaining a balance of red, orange, and yellow rays and evolve to higher levels whereas those with relatively balanced lower rays can exercise a certain control over their mind and move forward. When a person is not in control of his mind, it is difficult for him to keep his environment - including the physical body- under control, and, if given a free hand, can cause much suffering to himself and others.

Though there is a hierarchical nature and function among the rays, all rays are equally important as each one is indispensable for the functioning of the soul. As each lower ray becomes balanced/ cleared, the next ray can get more of the cosmic energy and our spirit can meet the incoming *prana* at a higher ray, taking the soul one step forward in its evolutionary journey. As human beings, we need to have our lower rays viz. red, orange, and yellow cleared/ balanced to a certain extent so that enough energy can move to the green-ray and we can work on this ray and ensure promotion or graduation to fourth density. The red ray needs serious attention as this is the first point of the influx of *prana* and any imbalance in this ray can reduce the overall energy flow causing imbalance and preventing further development of the soul.

The ultimate purpose of our life is the development or evolution of the soul or making spiritual progress and experiencing oneness with God in an ascending order until we merge with God. It is about learning, growing, and realizing our true nature in an ascending order. As long as the individual is struggling with survival issues, there is no possibility of spiritual progress. It is true that it is imperative for our material and emotional needs be met before we can launch ourselves in the pursuit of Spirituality, as Maslow had so aptly put that self-actualization can happen only after the more basic needs have been met. In other words, those who struggle with the world around them, and do not get the opportunity to look inward and see their true being, have little chance of graduating from third density to fourth density. Only when the individual can display selfless love and genuine compassion, he becomes ready for graduation to fourth density positive.

As in every other density, we have to live many lives as a human being in third density to learn our lessons and make our choices; and each of our choices (exercise of free will) determines whether we learn and progress to the next lesson or continue with the same lesson. To the extent our day-to-day activities become harmonious with the purpose of life (short term and long-term goals), to that extent we can experience peace and happiness.

Until we become self-conscious and learn our lessons, we may have to go through those experiences (not pleasant in many cases) that are designed to teach and evolve us, either in the current incarnation or in a future one.

When we begin our third-density evolutionary journey as a human being, we have limited awareness and when we reach a certain higher level of awareness, we are ready for graduation to fourth density. We initially learn from trial and error and as we evolve (become more aware/ intelligent), we learn to identify the cause and effect relationships and avoid certain actions that could cause pain and choose those that give pleasure. As we progress or evolve, our needs become higher and more refined. At the beginning of third density, we behave more like an animal and towards the end, we become like an angel.

The Law of Action or Karma ensures that everything will happen according to our inner state, ensuring that 'we reap what we sow'. Therefore, we must give or do to others only what we would like to experience in our life. Everyone and everything that shows up in our life is a reflection of our inner state. Those who resonate with our frequency move into our lives and those who do not resonate move out. This mechanism ensures that everyone gets those specific opportunities and environments needed to complete their requisite learning and move up in the scale of life. This Law also ensures that we do not try to jump or skip lessons and learn in a thorough and orderly fashion.

How to evolve faster

Understanding and perceiving our soul, which is our eternal self and true nature, is the first step towards mastering our life and evolving consciously. Soul's evolution is primarily guided by the inner tension which is caused by the urge or longing or will of the spirit, which resides in Indigo ray. The soul, specifically its spirit complex, is aware of its source and purpose of life and is constantly trying to find union/alignment with the Super soul until it loses its individuality and become one with It. This tension causes the soul to seek ways for evolving and reaching higher levels of conscious awareness. This tension is the basis for our numerous incarnations and learnings through varied lessons/experiences which do not go away until the soul reaches the seventh density and merges with the Creator. This tension makes the individual discontented with every achievement materially and spiritually and seek for newer/ higher ones as he intuitively knows there is more to learn/experience.

In each density, the methods and nature of evolution vary as each higher density experience is designed to take the soul one step closer to the ultimate goal of attaining complete union with the Creator. In first density, the tension is to build a foundation/ environment so that souls have enough opportunities and options to experience life in an ascending order and grow and move to second density. In second density, the tension is to use maximum opportunities for development as an individual and grow in awareness and progress to third density. In third density, the soul has become aware of itself and experiences the tension of realizing more of its true nature through social interactions/ relations and move to fourth density. In the fourth density, the tension is to live in perfect love in an open environment. In fifth density, the tension is to attain maximum wisdom. In sixth density, the tension is to balance love and wisdom and live as an individual with perfect alignment with the Creator. In seventh density, the tension is to lose the individuality and attain complete union with the Creator.

Evolution or transformation is a continuous and eternal process as higher awareness is possible only when each of the lower rays of the soul is developed and balanced by the experiences and knowledge gained from each density and through numerous incarnations. So, for a soul that has reached third density, survival and sexuality are the causes of tension in its red ray whereas personal relationships are the causes of tension in the orange ray. Individuality (ego) and social relationships are the causes of tension in the yellow ray and matters of love cause tension in the green ray. Longing for wisdom creates tension in the blue ray and self-realization causes tension in the indigo ray. Violet Ray does not have any particular tension as it just reflects the overall state of the soul. For each individual, the tension varies depending on the level of his development. At this point, most of humanity is struggling with the tensions in red, orange, and yellow rays. Many experience tensions in the green ray and a smaller number of experiences tension in the blue and indigo rays.

The more the imbalance, the higher will be the tension, and the lesser the imbalance, the lower will be the tension experienced by the soul. In other words, there will be more tension in a person who is farther from the goals of each ray (both incarnational and ultimate). Those who are stuck in lower rays struggle with their lower ray tensions because of their limited awareness as higher rays are not open for work (this also means they don't experience the tensions of higher rays), making them less equipped to manage the tension. They attempt to overcome this tension by suppressing or avoiding or running away from the reality of self/life by turning to various distractions like power, fame, money, objects, people, ideologies, organizations, religion, politics, etc., and trying to forget themselves through entertainment, sex, drugs, etc., with the false hope that these transient things will fill their emptiness and make them satisfied, happy and peaceful. What they are doing is to remain happy by managing the external environment. Though there is a temporary respite, these obsessions and addictions make the person more frustrated and restless by creating more problems and taking him/her further away from the true goal/purpose of life.

There will be considerable tension in a relationship where each individual is focusing on different rays.

If the focus of two souls is on the same ray they could adjust well and can have a stable relationship. As they are sharing similar tension, they may feel comfortable with each other as one will be able to accommodate and support the other. Only people with similar or compatible/complementary tensions can coexist in a close relationship peacefully. The more the gap between the focused ray of two souls, the wider will be the gap between them, and hence they could not enjoy a cordial relationship. This is the reason why negative people like to be in the company of other negative people and environment whereas positive people like to be in the company of other positive people and environment. It will always be easy to deal with people who have similar/compatible focus or are open to the possibility of raising their awareness to higher levels than dealing with those who think they have arrived and hence are stuck.

As the individual spiritually evolves and the vibration becomes more and more harmonious/balanced, his tension becomes lesser and easy to manage. A person whose lower rays are balanced has the capacity for increased awareness, clarity, and is capable of attracting and experiencing higher/ positive feelings because of the availability of higher rays for conscious work. As the positive feelings increase in the soul, to that extent the negative feelings get replaced and the whole being experiences harmony and peace. Therefore, what we primarily need is a balanced soul; once that happens, everything else will fall in place. Therefore, anything that helps an individual to achieve this balance should be encouraged.

A consciously evolving person can easily recognize how people and circumstances changed at each level of his life journey when he moved his focus from one ray to another. This change can be compared to the changes that happen to a computer system when the operating system is upgraded or replaced, making some of the earlier applications incompatible or obsolete It can also be compared to the changes that happen to our physical environment from one climatological season to another.

Negative thoughts/feelings distort and/or block the flow of energy within our soul which leads to a cascading negative effect on the soul and its environment. Therefore, we need to be watchful of those triggers or catalysts that cause negative feelings/emotions like fear, anger, frustration, sadness, jealousy, hatred, etc. They may come from people or circumstances and we may encounter them repeatedly. In most cases, whatever is the behavior/nature of these individuals/circumstances that bugs us or causes negative feelings/emotions in us, is an area with which we are still struggling. When we analyze or meditate on these catalysts then we can see that each of them is trying to teach us some virtue like patience, self-control, forgiveness, love, etc., which is the opposite of our imbalances.

Our feelings/emotions tell us whether we are in alignment with our true nature and purpose of life at any point in time. Positive feelings say we are in alignment and negative feelings say we are not in alignment. Negative feelings tell us to change whereas positive feelings tell us, it is where we need to be. These feelings/emotions guide us to be in alignment with our true nature and purpose. Further, our feelings/emotions determine our reality because we attract what we love, fear, hate, etc. If we are judgmental then we encounter judgmental people all around us. If we are nonjudgmental then we encounter nonjudgmental people in our day to day interactions. If we hate people/circumstances, then we come across lots of hateful people/circumstances and when we start loving others, we suddenly find a change in those people/circumstances towards us or we see those people/circumstances being replaced with loving ones. This means that we (specifically our inner state) create/ control our environment and circumstances. This also means that when we get transformed or balanced, then we do not need certain lessons anymore and they just go away, and we avoid attracting such imbalanced people or circumstances in our life again.

It is therefore important that we introspect or review our life regularly, at least once a day, to see where we need change or improvements. We should examine our lives to figure out what our external environment is trying to teach us by those circumstances and behaviors of other-selves and by the feelings they are creating in us. Furthermore, we should figure out why we did what we did and what were the mistakes we committed, and what would have been the ideal choice of action. Then we need to accept the day's experience as it is and forgive other-selves and our self for the mistakes, if any and thank every participant for his attempt to teach us. We should also bless all those people whom we perceive have wronged us or have caused inconvenience or pain to us and should love them unconditionally. With this, we will be able to balance or transform our soul from any blockages or imbalances.

To further simplify this process, we can divide it into four steps. The first step is to become conscious of the imbalances. The second step is to accept their existence within us. The third step is to forgive ourselves and others for these flaws and the last step is to develop/ manifest the opposite of those imbalances/feelings/emotions and express unconditional love. Once the balancing is achieved, we can see a corresponding change in our environment, and we do not encounter such catalysts (behaviors or situations) anymore. The balancing work is a continuous process as imbalances and blockages can happen any time due to the choices we make in our life. Self-analysis (introspection), meditation, intuition, and dreams are important tools we can use for clearing and balancing the energies of our soul and evolve.

We should meditate on those subjects which we want to know or on those we are struggling with, as and when required (it may not be necessary that we have to set apart a time and/or place for meditation if we can close our mind to the external world and focus on what we need when we need it). Through meditation, we get answers and/or solutions instantly or after some time, either directly or indirectly, through dreams or our environment. Meditation is our attempt of seeking answers and solutions and as we seek, surely, we will find. We could also use meditation to calm our mind by not feeding the mind with new thoughts and allowing it to find its natural balance. If we take time to reflect on our lives and take corrective actions, try to meditate, and think positively regularly, we can easily keep all our rays/*chakras* in a good condition and experience perfect health and happiness.

There is no point in trying to suppress any of our thoughts or feelings/emotions as they will not go away until they are acknowledged and released or replaced with other thoughts. What is suppressed or resisted will persist. Suppression and resistance will only prolong the problems and increase the miseries.

Further, we should start trusting and using our intuition faculty, awareness of which can guide us and keep us on the right path. If we follow our intuition, then we will find new or unexpected opportunities and will be able to make the right choices at the right time. We may not know the entire path to the destination or goal, but we will be guided to it by intuition, just like when we go to a new place, we don't know the entire route but there will be signboards or people at every intersection to guide us. A trained mind can make use of intuition in decision making and avoid possible mistakes. Just like proper planning helps us to simplify and complete our work faster, asking the right questions simplifies and makes our spiritual evolution faster.

Dream is an important tool, which I have personally found to be very useful, that can really hasten our evolution. Dreams are caused by the subconscious mind and the subject matters are relevant to our present life and chosen based on the balancing requirements of the soul. If we meditate or seek answers or solutions to certain questions or situations, then the dream life will be directed towards answering those questions or managing those situations. (A case in point is the famous discovery of the Benzene ring structure by August Kekule in his dream, wherein he saw a snake trying to bite its own tail).

In the dream state, our soul is experiencing life without our physical body. Dreams help us to experience certain things that we cannot or are unable to experience in our normal/waking life. In other words, dreams help us to experience life without the limitations of the physical body, which is always limited by time and space. In a dream state we can visit different places (including those of other dimensions), meet different people (from past, present, and future), engage in various activities that we love or hate or are afraid of doing, and gain much experience and knowledge and its resultant balancing/freedom.

All the experiences, whether in a waking state or sleep state, are part of our life and there is a reciprocal relationship between them, one affecting the other. Like every other life experience, dreams of a similar nature repeats until we learn the lesson. As we evolve, the contents and nature of our dreams also change. Though the catalysts from the dream experiences are mostly used during the dream itself, which is why we feel different after each dream, there will be some messages/lessons for our conscious understanding and use also. Therefore, to make maximum use of the dream life, we should try to recollect and meditate on the dreams we just had and figure out the messages/lessons they are trying to teach, as soon as we are awake.

The key requirement for using these tools is a calm and focused mind. A calm state can be achieved by removing distractions in our life and focus can be achieved by training/discipline. Yoga and meditation also help us to calm and discipline the mind. Both physical and mental health are important for a calm and focused mind. Since the health of our body and mind is dependent on the environment (both internal and external), we should practice natural living to keep it in control. The ideal state of the mind should be that, whenever required, we should be able to calm or silence it and focus. A person, who can use the aforesaid tools and forgive and accept self and other selves can consciously make his transformation faster.

Our objective is to become free, without any kind of bondage. Only we can make ourselves free by knowing the truth if we choose to do so. Expansion of awareness comes naturally in an ascending manner as we evolve in our life by being faithful to everything we have and do, and sincerely seeking the truth in everything. Others cannot make us free but some of them can help us to see the truth so that we can become free. It is always better to understand the truth ourselves rather than blindly believing someone's perspective because only when we know the truth ourselves, we can experience the freedom it brings. Till we consciously choose spiritual development as our primary goal, our spiritual development or evolution or transformation happens subconsciously and at a slower pace, often making the person repeat the lessons, during the current incarnation or the future ones.

Being self-motivated enough to live a focused life and consciously evolve is the most difficult thing for an ignorant person. Only an awakened person is self-motivated to live a focused spiritual life. Let us remember the truth that we are on an eternal journey of learning and development and there is no point in resisting change or blaming others. There is nothing in the universe that is not subject to evolution. Be flexible and open as that will not only support our spiritual development but also save us from many diseases of the mind and body.

As our awareness expands, to that extent we can understand our soul and see those triggers that make it comfortable or uncomfortable.

We need to approach everything in our life with an open and detached mind and allow what is not needed to just fall away from us. Detachment does not mean inaction. It means, though we are passionate about our actions, we are detached from the outcomes they bring. Getting attached to anything of the material world can surely slow down our spiritual progress. Drop all those things that do not serve us or are hampering our spiritual progress. Since whatever we think, do, or experience affects our soul and its environment, we must organize our lives in such a way that we can think positively, do good, and experience peace and happiness continuously.

If circumstances are putting pressure on a person, then it shows that he may not be on the right path as far as his incarnational goals are concerned. An ignorant person, by his negative actions, demands/invites pain into his life, which acts as a constraint and compels him to change the path. Similarly, an awakened person, with his positive actions, demands/invites peace and happiness into his life which then further motivates him to continue on the same path.

For people with low levels of awareness, it is better to follow the teachings of those who are more evolved and do what is logical or comfortable or feasible to them, as long as it does not cause harm to anyone. The touchstone of every action is whether it helps or harms another individual. In other words, whether we treat others as ourselves or as separate from us. Because when we see or treat people in a certain way, we also get treated similarly, as we reap what we sow because of the Law of Action.

As we consciously evolve, our vibration changes and we no longer resonate with many things that were important to us until then. We may try to withdraw or isolate ourselves from many mundane things to continue experiencing the connectedness and remain blissful. As our soul grows in awareness and oneness, it becomes aligned with its natural environment also. As we move up the scale of life, both our inner and external worlds become better and better until everything becomes so harmonious that we are ready to move to our next level of existence. Sometimes it may look like chaos all around but for the evolved souls there will be an inner sense of calmness that will help them to remain unaffected by the actions, thoughts, opinions, and beliefs of other selves.

Only an evolved person can see the truth behind all effects and know what is happening. He can see or sense more clearly the truth in everything including the intentions of people around him. As we progress in the scale of life, our awareness grows, and we get a better understanding of what we are doing and where we are going. As the soul evolves and the awareness expands, there will be definite and noticeable changes/experiences in the individual as the overall vibration of the soul becomes higher. Some of these benefits in a positively oriented individual are:

1) **Blissfulness** - able to experience love, oneness, freedom, gratitude, peace, and happiness continuously.

2) **Mindfulness** - increased awareness – even continuity of sleep and waking states (where dreams supplement the conscious life).

3) **Abundant life** - having a meaning and purpose to life.

4) **A powerful mind** - intelligent, intuitive, sensitive, discerning and positive.

5) **Balanced personality** – having self-control, confidence, contentment and detachment (not easily affected by circumstances).

6) **Increased health and wellbeing** of the body and slowing of the aging process.

7) **A sense of immortality** and loss of fear of death

****

Chapter V: The Law of Action or Karma

How our action return to us with a result that is equivalent to the 'quantum & vibration' of energy we utilized in the action. How this law affects us mentally and physically every moment of our life and how we can become spiritually free and live an abundant life.

Introduction

The universe is governed by Natural Laws which are the natural tendencies or principles of life or creation. Some of the fundamental natural tendencies are Vibration, Attraction, Repulsion, Polarity, Action, and Evolution. These principles apply to both physical and metaphysical worlds and operate at every level of creation, from a photon to galaxies.

The Law of 'Action' or 'Cause and Effect' or 'Karma' is the natural tendency of creation which restores or regains the balance or rhythm or harmony of the creation whenever it is distorted. 'Karma' is a Sanskrit word that means 'Action' but generally used in place of 'fate' or 'destiny' or 'give and take account'. As per this principle, for every action, be it benign or malignant, creative or destructive, there is an equal return or reaction. This principle ensures that 'as you do is done unto you' or 'as you sow so shall you reap'. Through the mechanism of the Law of Karma, the soul is purified and evolves continuously, until it finally merges with the Universal Creator.

As with every natural law, this is an impersonal law that works everywhere and all the time. All great masters were aware of this law and hence they taught to 'do unto others as you would have them do unto you'. Newton's third law of motion states that 'for every action, there is an equal and opposite reaction' which explains the operation of this principle as far as the physical world is concerned.

The Law of Action maintains the justice and order of the universe by giving a commensurate return for every action.

The Mechanism

At the fundamental level, everything is soul/photon which is a web of energy complexes. The mechanism of the Law of Action is that, every action whether a thought, word, or deed causes a certain change in the quantum and nature of the energy complexes of the soul. These changes get recorded or stored as a memory in the soul, which ensures that at some point in time, a corresponding quantum and vibration of energy flows back to the doer via another soul or circumstance.

This Law ensures that in every action the soul earns and later receives an equal return of peace and happiness or pain and suffering. Every positive thought or deed generates a 'merit' or a 'blessing' while every negative thought or deed generates a 'demerit' or a 'curse'. Whenever one does a good deed, it is bound to give a positive return and whenever one does an evil deed, it is bound to give a negative return. Each time we forgive and love those who create trouble for us or inflict pain on us, we are using the energy positively that ensures a similar positive return to us. Similarly, each time we hate, inflict pain on others or seek retribution on those who persecute us, each time we punish and refuse to forgive, we are using the energy negatively that gives a similar negative return to us.

Jesus said 'do not resist an evil person. If anyone slaps you on the right cheek, turn to them the other cheek also' (Mathew 5:39). This message forms the core of Jesus' teaching as he insisted on using energy positively through nonresistance and nonviolence until his death. The whole sermon on the mount is a sermon on the Law of Action as it explains the cause and effect mechanism at a deeper level in the context of the Old Testament principles. He was trying to put the limited application Old Testament principles to their full application or true perspective and thereby correcting many of the misconceptions prevalent in the society. No wonder, majority of his audience was shocked (and the majority is still shocked) to hear such higher truths as it contradicts the instinctive response of its lower nature.

All our actions do have a return that is equivalent to the 'quantum & vibration' of energy we utilized in the action. The return and equivalence are based on the principle that holds us responsible for the energy we use in our actions. Because of the operation of this principle, all the choices we make and actions we do return to us like the swing of a pendulum. The Law of Action assigns results according to the quantum and vibration of energy we use. It is simply an immutable law that gives appropriate results to every action irrespective of who did what. In other words, under the Law of Action, there is a return for every 'action' deeded by us each moment.

If we use energy negatively in our actions, no matter what the provocation or circumstance may be, we will be subjected to that same negativity in our lives. If the action is good, the return will also be good. Actions can be 'mental or physical' or 'direct or indirect'. Even actions by an established institution like a government or a business corporation, under the cover of a statutory rule or a business policy, do affect our individual Karma as long as we have approved their wrong actions or chosen to look the other way or have supported them in some way. This is the reason it is emphasized that we have to forgive and love everyone irrespective of what they have done to us. Nonviolence in thoughts, words, and deeds is essential if we have to escape bad karma and evolve spiritually.

As long as there is a debit or credit account, a balancing act needs to happen in due course because of the operation of this Law. Throughout our lives, we are either settling old accounts or creating new ones. When a settlement happens, the particular recording/ memory within the doer, that has attracted this energy, gets erased or released or balanced and one may say, the 'karma is dissolved or balanced'. If the account cannot be settled in one incarnation, it is carried forward to the next incarnation. Though in most cases we may not be consciously aware of the give-and-take accounts that happened in a previous life, by analyzing the experiences we are going through and the lessons we are learning, we can figure out the type of karma we had created.

The Law of Action also ensures that all our subsequent actions correspond to what we have already earned or become. Therefore, a negative person attracts more negative thoughts from other sources which intensify his feelings/emotions leading to mistakes or errors in judgment. This happens because, under the influence of strong negative feelings/emotions, the dark deed is perceived by the individual as being good and he tends to justify his actions. The soul subconsciously allows it to happen so as to suffer the consequences which will provide opportunities for learning a certain lesson and experiencing resultant transformation. The recording of the changes in the quantum and vibration of energy within our soul is the compelling force that guides our thoughts and actions.

Energy used negatively compels us to act foolishly and/ or hurtfully. This situation is aptly described by the Sanskrit saying '*Vinaash Kale, Vipreet Buddhi*', which means, in bad times one's intellect fails or acts foolishly. Whereas energy used positively compels us to act wisely and/or compassionately. Further, every time we do negative/ unkind acts, we increase our chances of being influenced and supported by negative souls. Whereas, every time we do positive/ kind acts, we increase our chances of being influenced and supported by positive souls. This support makes a lot of difference in our lives as it is the unseen hand that makes it easy or difficult to make or break our endeavors.

Because of the operation of the Law of Action, a person or society, by creating misery for others, invites misery into their own life. Similarly, any country that inflicts pain and suffering upon another country cannot escape the just consequence of its actions, when the time is ripe, when its cup of iniquity has become full. When darkness prevails over the majority of people on the planet or in a country or community, there will be mass sufferings, either through war or natural calamities or pandemic diseases. Generally, when an individual contract a disease or gets into painful situations, it is because of individual karma. Whereas a pandemic disease like Covid-19 or a world war is consequent to the karma on a planetary level. Negative thinking and aggressive and abusive behavior towards each other and nature creates disturbances in the energy vibrations of individuals and the planet and leads to events like diseases, wars, earthquakes, tsunamis, cyclones, floods, climate change, etc.

Unless there is a demand, no supply will be made. Demand calls for supply and every demand will find or attract a corresponding supply. A person experiences suffering or joy when his give and take account demands it. A corrupt society creates a demand for corrupt systems and corrupt leaders. A nation is ruled by corrupt and/or cruel people when the majority of the society demands it by its corrupt and/or cruel behavior. As long as the demand remains, attempts to change external conditions through political/ social revolutions or military/ economic interventions will not lead to any sustainable solution. Only when people transform themselves, the corrupt systems slowly give way to better ones and when the majority of people of a locality or country become positive, then they are fit to be ruled by wise, honest, and benevolent leaders. If we look around us, we can see perfect examples of this truth.

Our environment (surroundings and circumstances), especially people, acts as a mirror where we can see our reflection/image. They show us the mirror and help us to transform as we recognize many of our imbalances when we see them in others with whom we have close/regular interactions. This is all the more so with family members and other close relationships, who put the magnifying glass over our shortfalls or imbalances and help us to recognize and balance them. Like everything else, we attract such people and circumstances to help us correct our imbalances. Our incarnation and life are preplanned and re-planned in such a way that we come across those people and circumstances with whom there are karmic accounts to be settled. This is the way our soul plans its evolution in each incarnation.

We can see a repeating pattern of these reflections or projections at multiple situations in our life. And it appears that we are forced to endure those situations as if we are stuck there. In most cases, whatever is the behavior/nature of these individuals/circumstances that bugs us or causes negative feelings/emotions in us, is an area which we are still struggling to balance or master. These catalysts continue in our life until we fully evolve/transform. When we analyze or meditate on these catalysts then we can see that each of them is trying to teach us some virtue like patience, self-control, forgiveness, love, etc., which is the opposite of our struggles/imbalances.

There is a reciprocal relationship between our thoughts and the environment. Hence, we can see that when our thoughts change, our environment also changes and vice versa, from moment to moment. As we are one with the universe, we cannot harm others without harming ourselves and we cannot bless others without blessing ourselves. Harboring negative feelings against another person will invite a matching response from that person, even without him knowing it consciously. Similarly sending positive feelings to another person will cause a matching return from that person. If we are positive, our environment will reflect that state. If we are negative, our environment reflects that state. If we express love in all our acts, then over a period of time, we are cleansed of all negative feelings/emotional energy and are left with only positive vibrations within. And our environment, including the physical body, faithfully reflects our inner state –the state of our soul.

We are born into a particular family/community/country with certain circumstances that are conducive to undertake our learning and balancing requirements for our give-and-take account. As per the individual requirement, one may take up the body of a male, female, or transgender with a perfect or a disabled body. Two enemies or friends may come together in a future incarnation as husband and wife, parent and child, siblings, employer, and employee, etc. with unique circumstances so that they can settle/balance their accounts. Such close relationships and circumstances ensure that they cannot easily run away without working out their karma. Likewise, any unfulfilled promises or desires or commitments would also require two souls to come together in future incarnations to settle the account. So, it is better for every individual to improve his/her existing relationships wherever possible by giving unconditional love rather than trying to escape from it through separation, as debit/ credit accounts created in any incarnation will attract the involved individuals to engage with each other in future incarnations until such karmic accounts are balanced so that the energy within the soul is balanced.

Each incarnation is arranged in such a way that it provides a soul with appropriate circumstances to balance the 'give and take account' and receive the necessary lessons to be learned so that it can move to the next level. Nobody knows what the other person needs for his growth and in most cases, the individual may also not be consciously aware of his own needs because of limited access to the greater part of the mind (subconscious mind), while in a physical body. The Law of Karma ensures that every being will be at the exact place where it has earned the right to be, by all its previous thoughts, words, and actions, including those of the previous lives. The sum of the development of the soul (both from past incarnations and the present incarnation) and the present actions and reactions, both mental and physical, will determine our destiny. As our soul changes, our environment changes, our preferences or choices change, and our destiny changes.

Some people believe that they are 'predestined' by God or some other power to a certain fate and they have no escape from it. The truth is that we are not predestined to a certain fate by God or anybody else, but we are predestined by ourselves, by our actions. The memory or our actions accumulated over past lives and the present life constitutes the 'give and take account' or 'karma' that determines our 'destiny'. This is a self-made destiny and not 'predestined' by 'God' as believed by some. As per this Law, we will reap according to what we sow (according to our actions - thoughts, words, and deeds). It is the return or reward from Nature for any act, whether positive or negative. Many are unable to understand this concept because they have been taught that their 'destiny' lies in someone else's hands or it is just a matter of 'good luck' or 'bad luck' about which they can do nothing. This Law teaches us that there is no such thing as 'luck' and whatever happens to us is the result of our past actions (thoughts, words, or deeds).

This Law ensures that the aggregate of our past lives (of previous incarnations and the present life until now) determines our destiny. We can ignore the Law and suffer its consequences or consciously cooperate and enjoy peace and happiness. Until we are awake and attain a certain balance, we experience a roller coaster life because of the operation of the Law of Action. It is an immutable Law and we cannot wish away the consequences of our actions. Nobody can escape this law - we are paid in the same coin that we give. This law helps us to be reminded that our environment and our situations today are the net result of what we thought and did in the past.

Everyone will receive according to what they give in thought, word, or action. And the receiving is faster now because our planet has moved into higher vibrations of its next phase of existence, that is fourth density. Though many do not understand the underlying mechanism of this principle, everyone, by experience has learned the relationship between cause and effect and hence avoids doing certain things that could give them unpleasant results, just like the child who attempted to touch the flame and burnt its fingers tries to keep a safe distance from it next time. This law does not put us in a hopeless situation, but it teaches us what is good and what is harmful.

Being positive and doing good is emphasized by every master because one cannot escape the reality of the Law of Action and must reap what is sown, both physically and mentally. The physical reaping of experiences and circumstances happens while the soul has incarnated in a physical body whereas the majority of the mental reaping happens while the soul has left the body and is waiting for the next incarnation in the sub-density or metaphysical realm. During the waiting period, the soul has to go through a life review when it has to feel all those feelings that it gave to other souls either directly or indirectly by its thoughts, words, and actions during the earthly life. The intensity of the feelings will be higher because there is no physical body to share the impact. This is the state of hell for those who inflicted pain upon others and heaven to those who made others happy. The majority of the mental reaping can happen in the metaphysical realm only because experiencing even a fraction of those multitudes of emotions could make the person insane or invalid while in a physical body.

How to consciously use/align with this Law

We create our lives and we can only create those things that are in alignment with our inner state. So, our world will be based on what we love and/or what we fear/ hate. Today's circumstances are the result of our past choices and actions and future circumstances will be the result of current choices and actions. Further, our past and current choices also determine what type of choices will be available to us in future. It may be true that we may not be able to pinpoint the 'ultimate' cause of most of the events in our life as every event is an effect of a previous cause, and that previous cause is also an effect of a cause still more remote. At the back of every action, we find some actions that lead us to the one that we may be considering now, but we cannot easily recognize what triggered those initial actions. However, a conscious person can see a pattern emerging from these events which correspond to certain behaviors or beliefs or thoughts.

Everyone has the freedom to choose whatever he or she likes but cannot escape the consequences of the choice. Sufferings and joy are not arbitrarily bestowed upon us by any external power. They are the consequence of our past choices only. However, the power to exercise this choice is directly proportionate to the soul's development. The soul's development manifests as character and personality, which consists of the wisdom, knowledge, beliefs, and biases it had accumulated from the experiences of its evolutionary journey. The more evolved the soul, the easier for it to make the right choice or judgment. Only an evolved/ matured individual can consciously exercise the right choices under difficult circumstances. Impulsive behaviors show the inability of the individual to make a conscious choice.

In the physical world, we know the result of every action can be manipulated with a counteraction, and hence two equal and opposite forces can neutralize each other. Similarly, adverse circumstances can be changed to favorable ones when we take responsibility for everything that happens in our lives and our circumstances, stop blaming others and forgive and love others irrespective of the situation. This is how we can stop the wheel of karma or cycle of creating more negative patterns and allow our souls to be cleansed of the accumulated imbalance. This way we can become spiritually free and save ourselves from accumulating more bad karma. This is possible for everyone who is consciously working towards spiritual development. However, for the ignorant person, a change in destiny happens for better or worse at a slower pace based on his routine thoughts and actions, without any conscious effort to change it, as he is yet to become aware of such a possibility.

As a human being, each of our incarnations is an opportunity to make rapid spiritual progress and get closer to our final goal i.e. achieving absolute union with God. Each one of us is responsible for our happiness or misery and we can halt the wheel of suffering by transforming our mind by knowing the true nature and mechanism of life and then living in harmony with Nature. When we take responsibility for our lives and consciously work towards clearing the debts and create credits then we can surely change our destiny to a happy life.

The nature of our soul is constantly being changed by our thoughts and feelings/emotions which in turn influence and change our surrounding environment. Every benevolent action helps our soul to become positive and increases its light quotient and awareness. Every malevolent action makes the soul negative and decreases its light quotient and awareness. When we attain a certain balance within our rays/ mind, we can see a positive change in all areas of our life. Therefore, the only sustainable solution to all the issues of humanity is to educate and awaken people to their true nature.

Do not try to forcefully impose change. Let the change come naturally when people and the environment are ready. We can only hasten its coming by knowing the truth ourselves and helping and supporting others to know the truth. This is the reason Jesus said, 'seek first the kingdom of God, and His righteousness; and all these things shall be added unto you' ('kingdom of God' is a reference to higher levels of awareness). We can surely accelerate our spiritual development and experience peace and happiness by unconditionally helping others to become free, happy and peaceful because of the operation of this cosmic law that ensures everyone receives according to what they give or reap what they sow.

> ***Note:*** *Until we have at least 4-5% awakened and evolved people, it is difficult to have a cleaner planet with kind and honest people as rulers.*

****

Chapter VI: Natural Living

All is One and hence our environment – both internal and external – influences and impacts us every moment of our life. How we can live naturally or live in alignment with the natural processes, laws, and flow of life and experience a healthy, happy, and peaceful life?

Introduction

Our planet is probably one of the most beautiful and habitable planets that have all the ingredients for supporting so many varieties of biological life. This supporting nature of the planet has provided us with everything to live a happy and comfortable life. All the conflict and suffering of the world are not because our planet lacks anything but because humans interfere and disturb the natural processes of life or work against the laws that govern Nature.

For physical or biological life to exist, an energy exchange system or ecosystem is essential. Suppose there were not enough light, air, water, certain minerals, microbes, and tolerable climate available on our planet. Then everything including the plants would die. If all the plants died, then herbivores that depend on them would also die. If plants and herbivores died, then the omnivores and carnivores that depend on them would also die and the planet would become unhabitable and desolate like our neighbor Mars. The present economic growth and development has come to us at a huge environmental cost and is leading us towards the possibility of an environmental disaster like Mars. We cannot deny the fact that with higher intellect, knowledge, and technology, humans can disproportionately impact the ecosystem compared to other creatures who share this planet with us.

Reckless commercialism and exploitation of natural resources by spiritually ignorant people have led to serious damage to the physical environment of our planet. Majority of people take nature for granted because they think that the planet or the environment has been there for quite some time in this condition and will continue to do so forever irrespective of what they think or do. And most people consciously do not do anything to support the environment or nature but rather destroy it with their ignorant ways of life. Though considerable damage to the ecosystem has already been done and is still being done for accumulating 'wealth' for someone or 'comforts' for somebody, the scarier thing is the misuse of technology and resources for creating highly destructive weapons like nuclear bombs and the possibility of a nuclear war.

Also, some spiritually ignorant people exploit and oppress others for money and power, which is the primary reason for poverty, inequality, and suffering on this planet. Religion is also used by some of them as a convenient tool to enslave and control the masses. It is also true that spiritually ignorant followers violently defend and promote their beliefs and masters without even bothering to check what they believe in and whether that benefits themselves or their children. We as humanity have reached an unsustainable position from where we have only two choices - either turn back and survive or continue and perish before the allotted time. One may wonder, knowing the fundamental interdependence of every creation, how could humans, who are supposed to be more intelligent and conscious than other species on this planet, continue with their suicidal behavior.

All of us indeed aspire to live a comfortable life on this planet and there is no doubt we all equally deserve it. However, it is not possible to live a happy and peaceful life when our fellow humans and Nature are suffering, as every act, we commit against Nature and each other will be returned to us because of the Law of Action or Karma. Since our environment – both internal and external – influences and impacts us every moment of our life, we need to live naturally or live in alignment with the natural processes, laws, and flow of life to experience a happy and peaceful life.

Natural living is about understanding 'who we are' - our true nature, both physical and metaphysical - and then consciously living in harmony with our environment. It is about being true to ourselves and living in the present with responsibility. Natural living is harmonious and will bring a balance in all areas of our life. Natural living practiced by even a small percentage of the population can definitely make life on our planet peaceful, harmonious, and sustainable. Natural living is the only sustainable solution available to us individually and collectively, locally and globally. An understanding of our metaphysical and physical environments and the fundamental principles of Nature can help us attain the right attitude and live naturally.

The Mechanism

The universe is multi-dimensional or exists in multiple densities and there is a hierarchical structure and function at every level of its manifestation. The material or physical universe that we can perceive with our five senses is the first, second and third-density physical manifestation of life, and there exist many other dimensions/sub-dimensions, both physical and metaphysical, that are beyond our perception while inhabiting a human body. Further, there is a fundamental interdependence and reciprocal nature in the universe and at every level of creation, the higher levels always incorporate and support the lower levels. Every creation supports other creations, directly and indirectly, each moment of its life. Within our solar system, Sun caused the creation of our planet and supports its survival and evolution, and planet Earth, with the help of the Sun supports each of its inhabitants to live and experience life. Within the planet, air, water, minerals, unicellular organisms, plants, animals, and humans help each other to create an ecosystem for experiencing life.

Every creation plays a definite and important role on the planet at each stage of its evolution, consciously or subconsciously and all the parts of nature work together to make a balanced ecosystem on the planet. Biological life is possible on planet earth because of the availability of light, air, water, and minerals in the right proportion.

Plants convert these elements into primary building blocks for physical life and provide food, shelter, and shade for other organisms such as animals and humans. Animals and humans in turn contribute to the life of plants, by providing carbon dioxide and helping in reproduction and decomposition of matter.

It is also true that we humans support each other whether we are conscious of it or not. All the services or products that we use every day are the result of the labor of another person. We eat because somebody is working in the field and we sleep because somebody is there guarding our locality and borders. Can we imagine a world where we are alone and struggling to just survive? Let us not forget, it is because of the fellow humans that we can live a safe and comfortable life. We should be grateful to everyone and everything for caring and supporting us every moment of our lives.

We are interconnected with every other creation, whether animate or inanimate and our thoughts can influence everyone/everything, positively or negatively. Our environment reflects our thoughts and there is a reciprocal relationship between our thoughts and our environment. Therefore, we can see that when our thoughts change, our environment also changes, and vice versa, from moment to moment. Harboring negative feelings against another person will invite a matching response from that person as well as from Nature in general. Similarly, sending positive feelings to another person will cause a matching return from that person and Nature. Therefore, we cannot harm others without harming ourselves and we cannot help others without helping ourselves. This is because of the Law of Karma and therefore, we should try to do good and make others comfortable and happy so that we can also get the same thing back.

The soul, specifically the mind, controls, and influences every aspect of the physical body and, therefore, a healthy mind is essential for a healthy body. Our life experiences are the results of continuous interactions of the soul, physical body, and the physical and metaphysical environment. Every thought makes changes in the energies of certain ray(s) and in the connected areas of the mind, which in turn affects the environment, starting with the physical body. The endocrine system under the control of the nervous system helps in coordinating or guiding the activities of our body. It accomplishes this by creating and releasing hormones through the glands which act as messengers as well as influencers within the body. Scientific research has proven that continuous stress is the primary cause for many of the so-called lifestyle diseases and a calm and happy mind could prevent as well as heal the body from such diseases.

The body is incorporating or absorbing every mental and physical activity of the soul and body respectively by creating a corresponding organic chemical (hormone) in the body. Some hormones take care of the physiological requirements of the body whereas some take care of the emotional requirements of the body. Therefore, the body's comfort is mostly dependent on hormones which are primarily dependent on the mind. Since hormones alter the physical state of the body, they can also influence the mind and accentuate its bias because of the reciprocal relationship between the body and the mind. Once a certain number of hormones are created in the body, they take control of the individual and compel the individual to behave in such and such way as per the nature of the hormones. Since some of the hormones are only for emergency/survival purposes, their continuous presence or higher levels in the body is injurious to physical and mental health.

Positive environment and thoughts cause happy hormones like endorphins, dopamine, oxytocin, serotonin, etc. in the respective glands which then spread that emotion all over the body. This is a cellular way of sharing happiness and wellbeing – a kind of organic display of love and affection within. In certain medical conditions, hormone therapies may take care of certain imbalances in the body but like many other medicines, they are not a sustainable solution. Whereas mental states can naturally manage hormones and we can consciously use the mind for changing the condition of the body. Therefore, taking care of the soul (unblocking and balancing the rays) and the external environment should be our focus for balancing or managing hormones rather than going for artificial hormone therapy.

Trying to engage in pleasurable activities and keeping the mind calm, positive, and happy can increase the levels of happy hormones in the body and keep us in a blissful state. Some of the activities we can undertake to increase these hormones are giving and receiving affection, showing kindness, love, respect, giving rewards and engaging in physical work, exercise, entertainment, sex, eating good food, taking bath, doing meditation, etc. Regular physical activity or exercise could not only improve our physical health but also help in shifting the mind to a relaxed and happy state because of the good feeling we get from the activity and the distraction it provides from possible tensions or worries (proper and sufficient use of the physical body is a must for our physical and mental wellbeing). While many of the physical activities give only short-term happiness, an evolved state of the soul can guarantee long term results. This is the reason why spiritually evolved people can experience a blissful state continuously without engaging in any such physical activity.

Yoga is a holistic system that can help both body and mind and support us to spiritually evolve. For spiritual development, one may focus more on techniques of breathing, concentration, relaxation, and meditation taught under this system. It helps if we start our day with half an hour of yoga and then consciously try to maintain a positive state of mind throughout the day. Those who are conscious and can organize their lives in a disciplined way can easily follow natural living and experience consequent health and happiness. Those who are under medication for various reasons need to heal themselves and get rid of those chemicals first to fully practice natural living.

We need to avoid those chemicals and environments that damage the glands, disrupt hormone production, confuse hormones, and interfere with the natural response mechanism, especially the immunity and healing system of the body. For that matter, we should avoid all kinds of pollution to keep ourselves natural, sensitive, and comfortable. When the environment is not comfortable, the mind can easily become agitated and perpetuate the negativity and the person finds himself in a vicious circle because of this reciprocal nature. Therefore, we should try to live with Nature and make physical activities and meditation part of our routine. Natural eating, meditation, sufficient sleep, and avoiding negative people and circumstances also help us to remain calm and peaceful. We need a healthy physical and mental environment for our physical, social, and mental wellbeing and even survival, and natural living is the only sustainable solution available to us. Natural living can help our body to work in a balanced way with its natural homeostatic mechanisms and prevent diseases. By understanding the role of hormones in our life and how they are being produced in our body, we can consciously create or increase those hormones that make us feel better and healthier, both mentally and physically. In like fashion, we can consciously limit or inhibit the release of harmful hormones in the body.

As an individual progress spiritually, his mind becomes calmer and clearer and subsequently, his sensitivity increases as the mind is more clearly able to reflect its environment. A sensitive person becomes more concerned or conscious of his health, environment, and self-development. He finds chemically treated food not appealing to his palate, crowded urban environment suffocating, polluted air and water disturbing, corruption and negativity intolerable, and so on. He also realizes that many of those things that he enjoyed earlier are no more making any sense – and there has been a major shift in his preferences or likings. For an imbalanced person, sensitivity can cause mental diseases but for a balanced person, it works as a warning mechanism so that we can work on issues on a timely basis. If we do not take corrective actions based on these warnings, then we may find ourselves in deep trouble. Therefore, it is essential that we understand these messages and take care of our environment, both inner and outer so that we can live a healthy and happy life.

What we need is an open and supporting environment to keep our mind calm and positive so that we could make our evolution smoother and faster. Therefore, we should try to live a simple life closer to Nature and spiritual people, as far as possible. We must consider all aspects of our life or have a holistic view and live in harmony with the universe or Nature as everything in the universe is connected to everything else and what we do affects us as well as everything around us. As we evolve, we will increasingly realize the oneness of this universe and will be able to live more naturally in a loving, compassionate, and responsible way. When we achieve a certain vibrational frequency and balance within our soul, then we can consciously cooperate and flow with the natural mechanism of life.

Those who are awakened or aware of their true nature will be able to live in harmony with their environment, both internal and external. They can live consciously, following the messages from Nature. Only such people can see the unity of 'All Life' and accept humanity as one. Such people assume responsibility and stewardship in all areas of their life and also in the society and environment around them. They manifest true love and bother helping themselves and fellow beings by taking care of the society and environment with activities like environmental protection, poverty alleviation, natural farming, natural healing, sustainable and economical activities like localization of manufacturing, renewable and free energy solutions, resource conservation through innovative products and services, value-based education, etc.

As each person is at a certain stage of the evolutionary journey, we cannot expect all of them to be awakened at this point. It, therefore, becomes the responsibility of the awakened individuals to take the lead and join hands with other such individuals to create awareness and support the masses to come up in life. Other ideas may be attempted, but ultimately it will be discovered that 'Natural Living' is the only sustainable solution that can bring peace and harmony to the individual as well as to the planet and ensure continuity of biological life on planet Earth.

> ***Note:*** *Some 'developed' countries pretend that they have excused themselves by outsourcing and exporting their pollution to other 'less developed' countries. But the truth is that both of them have to pay the price – it returns to them in some way or the other. It is a matter of great concern that most of these 'less developed' countries are fast catching up or overtaking their 'developed' cousins in polluting the planet.*

Natural Living for Wellness

Though our body is a single entity, it can also be considered as an intelligent biological ecosystem because it is made up of trillions of cells and microbes, each having a certain level of intelligence, with various activities happening simultaneously. Each of our cells goes through the basic processes of life like birth, movement, growth, response, metabolism, excretion, reproduction, and death like any other organism. Every day, billions of cells die and are replaced with new cells, within a healthy human body. Every part of our body from hair to toe replaces itself within a year, on an average, except in the case of bones and brain, which have a longer life span. The body also has an immune system that protects it from antigens and heals the body from various diseases and injuries. This is the kind of rejuvenation or regeneration, protection, and healing process our body is capable of under the control of the soul, which continues until the soul remains with the body.

Health is the natural state of the body and diseases are caused when the natural functions of the body are interfered/disturbed/distorted by any harmful substances, organisms, activity, or thought. A living body has the power and means to protect itself from any diseases or to heal itself, in case of an already diseased condition. Healing is natural to us and happens automatically through our immune system. Our natural immunity is dependent on several factors, such as the physiological condition, state of mind, quality of air, water, and food, personal habits, medicines, environment, etc. If we allow the body to regain its natural state - by slightly modifying the thoughts, habits, lifestyle, and also providing the right environment, the body can heal itself from any disease. In most cases, treatment is about providing a conducive environment and support like the right food, rest, peace, positive thinking, etc. to help the body in its self-healing process. A living body will always attempt to come out of the diseased condition using every resource at its disposal.

Neither doctors nor medicines do any healing but their services, if administered correctly, can support the body to heal itself. In the event of an accident, one may need medical support to get the wounds cleaned and stitched or the broken bones put in their place and to guard the wounds against possible infection so that the body can heal the wounds and/or join the bones. Therefore, it is of utmost importance that we keep our body processes in their natural state by not causing any disturbance or interference - whether by harmful chemicals, wrong habits, or negative thoughts - and support in the self-preservation and healing mechanism of the body.

Since the soul controls the body, we should consider the role of the soul also while supporting or trying to treat the body. A mechanical approach, specifically figuring out the chemistry without considering the power that manages all these processes within the body, is fatally wrong and has caused so much avoidable suffering already. Technology should be used to supplement and support our natural healing process and not to subvert it. We need to keep this in mind while taking help from any kind of medical support.

Modern medicine, which considers only the parameters of the physical body, for evaluation and treatment of various illnesses, cannot effectively manage our health and wellbeing, whereas holistic systems like the Ayurveda can surely help. During the Covid-19 pandemic in 2020 we have seen that the infection and mortality rates were disproportionately high in many 'developed countries', despite having advanced health care and medical infrastructure, which is a clear indication that the more chemicals (modern medicines) we ingest and the more mechanical our lives become, the less our natural immunity will be.

There is a direct reciprocal relationship between the body and the mind. A healthy mind impacts the body positively whereas an unhealthy mind impacts negatively and vice versa. Therefore, a healthy body and mind are essential for a happy, peaceful, and purposeful life. An unhealthy person will struggle to maintain the physical life and will be left with hardly any time or mind power to think about or attempt self-development. It is high time that humanity returns from the artificial habits acquired from the wrong culture/teaching/learning/ over the past few generations to the natural methods of living – a return to Nature.

Food is essential for our physical survival and is an important part of our daily routine. The physical body needs food for energy and nutrition as it is the fuel to sustain and maintain it. Without eating and drinking, we cannot survive for more than a few days. A person who eats and drinks properly, keeps regular physical activity or exercise, and keeps the mind calm and happy does not have to suffer from any disease. Even making small adjustments in our eating and drinking choices and habits could make a major difference in our health. With abundant physical health and mental wellbeing, one can truly work towards self-development and make strong spiritual progress.

The following section is an attempt to explain natural living with respect to our eating and drinking. It is about understanding the mechanism of our digestive system and the impact of our habits, preferences, and condition of the mind on this fundamental natural process of the body. It is specifically about how ingesting the right food and water and cooperating with the natural process of the digestive system or not interfering with it can avoid many physical problems and help us to live a healthy life.

Digestive system

The main organs of our digestive system are the mouth, esophagus, stomach, small intestine, large intestine, rectum, and anus, which everyone is familiar with. The basic physical and chemical processes of the digestive system are mastication, digestion, assimilation/ absorption, and excretion (expulsion of waste), each of them is explained below:

Mastication

Mastication or chewing is performed by the mouth with the help of teeth and tongue. Mastication breaks up the food into small particles and saturates the food with the saliva that is poured into it by the salivary glands of the mouth. Saliva contains an enzyme that begins the breakup of starch into smaller molecules. Mastication prepares the food for digestion. Properly masticated food becomes so smooth that it automatically moves into the stomach through the food pipe. If the food is not properly masticated, then it does not get properly digested in the stomach and turns into a waste and toxic material and damages the body.

Digestion

Once the food reaches the stomach, it comes under the involuntary control of the autonomic nervous system. The stomach performs three mechanical tasks. Firstly, it stores the swallowed masticated mixture of food. Secondly, the glands in the stomach wall (mucosa) add gastric acid and few other enzymes to the mixture and combined with its churning activity dissolve the solid contents.

Thirdly, it slowly empties the partially digested contents into the small intestine after assimilating the liquid portion. In about thirty minutes, this mixture (chyme) slowly leaves the stomach into the small intestine for further digestion and assimilation.

In the small intestine, enzymes from the pancreas, liver, and the glands in the wall of the small intestine mix with the chyme and help in the further breakdown of the carbohydrates, fats, and proteins in it. It then gets pushed forward for assimilation mainly in the small intestine and partly in the large intestine. Bacteria in the large intestine also aid in breaking down the waste material to extract the remaining nutrients from them.

Assimilation

Assimilation is the process of digested food or the nutrients and water becoming part of the blood. Assimilation mostly happens in the small intestine; however, the fluid portion of the mixture is taken up by the absorbents of the stomach and carried to the blood while the stomach churns the solid portion of the mixture. The large intestine is also involved in the assimilation process where water and certain vitamins are taken up by the system before feces leave the body. This means when we eat a toffee, that toffee enters our blood like when a drop of ink is added to water, it disperses into the water and becomes part of the water. Once the digested toffee becomes part of blood, it is then available to every cell of our body and cells use it for their organic processes as their maintenance and reproduction are dependent on the nutrients in the blood. Blood is the lifeline of the body and the quality of blood is according to the food we eat, the way we eat, and the way it gets digested and assimilated. So, we need to be mindful of what we eat and drink as that initially become a part of our blood and then become our body itself.

Excretion

Excretion is the process of expelling waste material from the body. After assimilation, the undigested portions of the food and bacteria remain in the large intestine and get expelled through the bowel movement action.

How to support/align with the digestive process/system

Let us now examine the natural living practices that can help us to make the best out of the digestive process and system.

Eating/Drinking Habits

Eating and drinking habits play a major role in the digestive process. Best practices we should follow while eating and drinking are explained below:

1) Mastication

Nature has designed and provided everything with a definite purpose and when we put it to its intended purpose, we can make the most out of it. The primary purpose of our mouth is for eating and drinking and we should masticate the food thoroughly making good use of the jaws, teeth, tongue, and saliva. Chew each bite for the maximum possible time, until it melts away with the saliva (in other words we have to drink the food). Take small bites and eat slowly so that food can spend more time in the mouth and mix with more saliva. Keep the lips closed while masticating for maximum results. While taking liquids, sip small quantities slowly.

Those with teeth issues could make use of their fingers or other support to make the food soft enough to masticate. Enjoy the taste of the food so that more saliva is produced by the salivary glands.

When the digestive system is functioning smoothly, there will be enough saliva in the mouth to keep it moistened and healthy all the time.

While masticating, solid food mixed with saliva initially turns into a paste and then semi-liquid form before automatically moving into the food pipe/ stomach. Therefore, we do not normally feel thirsty while eating, and at least for the next thirty to forty-five minutes after eating. Try to masticate slowly and allow the food to gradually melt away, rather than making a deliberate attempt to swallow.

2) Timing and quantity

We feel hungry when our body, specifically our cells, demand nutrients, and we feel thirsty when our body/cells demand water. Hunger and thirst are the signals from the body that it needs food and water respectively, and the body, particularly the digestive system, is ready to take/process them. This also means that eating when the body is not ready will lead to indigestion and other complications. It is also true that we give full attention to eating and enjoy the food when we are hungry. So, we have to eat when we feel hungry and drink when we feel thirsty.

A healthy person following natural eating won't feel hungry for the next 3-4 hours after a full meal as his/her body gets enough nutrition from that meal. This is more in the case of a person who eats raw vegetables and fruits as a meal as such foods give maximum nutrition. Whereas a diabetic person feels hungry very frequently because his/her cells are not able to accept/use the nutrients offered and hence crave/demand for the same all the time.

We need to maintain a minimum gap of 3-4 hours between each meal. Three meals a day are more than enough for a healthy person. This time gap will ensure that the body digests and assimilates the earlier meal before a new meal is taken. Also, it gives the body enough time to focus on other important functions, such as repair and maintenance, rather than using all the energy and attention towards digestion, assimilation, and excretion.

Further, we should not drag the eating session with any other activities like talking or using mobile, etc. Close the eating in a time-bound manner – the total time taken should be equal to the time required for proper mastication. There is no reason why a normal person needs more than 30 minutes to complete a meal.

Avoid drinking water just before, during, and immediately after eating food (though while eating dry foods we may take some liquid other than water). This is because water could interfere with the digestion as it dilutes the digestive juices and make them less effective. It is advisable to drink water ~30 minutes before and ~45 minutes after having food. Those who cannot follow this strictly, for various reasons, may take other liquids in small quantities. At other times one should drink enough water. While drinking we should not gulp quickly, rather sip it slowly in a relaxed manner (in other words we have to eat the water) so that it gets the opportunity and time to get mixed with saliva and becomes healthier and soothing to the digestive system.

Practice moderation – listen to the body and stop eating and drinking when you feel it is enough – don't eat until you are too full and unable to take any more. A person with minimum physical or mental activity may need two-three small meals but a person who does hard physical or mental labor will need more food. The quantity of food and water needed by a person will vary based on the age, body weight, physical and mental condition, nature of the job, climate, type of food, etc.

Also, avoid taking a bath for half an hour before and two hours after the meals. This is because a bath instantly brings down the normal temperature of the body making the body to commit its resources towards restoring the temperature. This adversely impacts the normal functioning of the digestive system.

Dietary Preferences

Every organism becomes the food of another organism once its controlling life force (soul) is separated or when another organism (normally from a different species) takes it up partially or fully as food. This is the order and nature of the universe- the bodies or produce of the organisms become the fuel or raw material for the development and maintenance of the bodies of other organisms.

We may eat any edible substance that Nature has provided us as food. However, if one has created biases against any food item due to religious or any other reason, then it is better to avoid such items because eating them may create emotions of guilt or fear which would not allow the food to be properly digested and upset the digestive system. Best practices in choosing the right type of food and water are explained below:

1. Varieties of Natural & Fresh Food

We need to eat a variety of food to ensure that our body gets all the nutrients it requires. Fresh fruits and vegetables are the healthiest types of food for a human being, followed by grains and animal products. One should eat what is comfortable and affordable and try to choose from the best locally available seasonal foods and include as much variety as possible. We should specifically include foods rich in antioxidants, vitamins, minerals, and fiber. Raw food, specifically fruits, and vegetables should be eaten for maximum benefits as it digests fast and provides more nutrients compared to cooked food, as some minerals and vitamins are lost during cooking. As raw food is more nutritious, eating it will avoid unnecessary craving for food or overeating as the cells/body do not demand more when all its needs are fulfilled for the next 3-4 hours by what we have eaten (this itself could take care of obesity and many other health issues). Eating raw vegetables and fruits, because of their high-water content, also fulfills some of the water requirement of the body.

Some foods grown at higher altitudes warm the body whereas some foods grown at lower altitudes cool the body. Therefore, it makes sense to use locally available seasonal foods as they help the individual to adapt to the climate of a particular area. Eating natural and tasty foods also helps in the increased production of saliva and other digestive juices and thereby improving the digestion process. Vegetables and fruits ensure that there will be enough fiber content available so that bowel movement and waste disposal become efficient. We need to restrict the use of refined sugar, salt, and simple carbohydrates and try to use natural salt, natural sugars, and whole-grain products whenever possible.

The condition of the food is an important factor and we should ensure, as far as possible, that it is free from toxic materials. Harmful chemicals, when ingested, interfere with the nervous system/hormones and damage organs and also cause destruction and imbalance in the friendly microbial communities (specifically in the intestine) which have a crucial role in the digestion of food and overall wellbeing of the body.

2. Natural Potable Water

Equally or more important than food is the water we drink as several health issues are associated with the wrong water or less water. Natural clean spring water is the best, but it is not easily available everywhere, especially in the cities. So the next best thing is to drink available potable water filtered using any normal filter. Storing water in a silver or copper vessel makes it healthier. It is better to avoid prolonged use of boiled and over-treated water.

Condition of Mind

As we have already discussed, the mind controls and manages the body with the help of the nervous system. And nervous communication is accomplished through neural signaling (radiation) where cells directly communicate with other cells and/ or by making glands to release hormones into the internal circulation/ blood, which in turn influence the functioning of the body. With every thought or feeling/emotion, there will be specific events/changes happening within the body which could be either healthy or unhealthy depending on the type of thoughts/ emotions. Negative (destructive) thoughts affect negatively and decrease the health and harmony of the body whereas positive (creative) thoughts affect positively and increase health and harmony.

1) Maintaining a calm state

While eating, we should try to be as calm as possible. We need to avoid all kinds of distractions while eating because only when the mind is calm, it can guide the nervous system and its linked glands to organize appropriate neurotransmitters/ hormones and ensure proper working of the digestive system. Try to remain in this calm state for at least 30 minutes after the meals. A calm mind helps the digestive system to function at its optimum level leading to proper mastication, digestion, assimilation, and elimination. Also, it is better to wait rather than eat when the mood is not right.

It could be beneficial to eat with a feeling of gratitude by thanking Nature in general and other humans in particular for producing and providing us with the food.

2) Focus and enjoy eating

While eating we should focus our entire attention on eating and enjoy the food. If the mind is focused on something else while eating, then the same will be passed on to the nervous system/glands leading to the creation of hormones unrelated to digestion. This causes two issues for the body in general and digestive system in particular. On the one hand, it does not produce the required hormones/enzymes, or the produced quantity is insufficient. On the other hand, the body has to deal with those other hormones, leading to confusion and malfunctioning of the digestive system. Therefore, we must avoid all distractions like talking, reading, watching television, and using phones, computers, tablets, etc., while eating.

Summary

The below table summarizes the relative importance and effects of each of the components of a natural food habit:

Constituents	Effect Percentage
Eating/Drinking Habits	**30%**
Mastication	10%
Timing and quantity	20%
Dietary Preferences	**40%**
Varieties of natural fresh food	30%
Natural potable water	10%
Condition of Mind	**30%**
Maintaining a calm state	20%
Focus and enjoy eating	10%

Notes:

1) As with everything else, at every stage of evolution our food habits also undergo a change and an evolved person may prefer small quantities of raw foods like fruits and vegetables instead of heavy meals.

2) Sometimes we experience hiccups which is a message from the body that there is some confusion going on within the nervous/endocrine system.

****

Chapter VII: Planetary Changes

The purpose and process of the ongoing transformation of our planet. The Impact of these changes on us individually and collectively and how we can make the transition smoother and faster.

Introduction

Chapter IV explained the evolution of souls and traced our evolutionary journey. As explained, our planet and humans are currently in the process of transitioning to the next stage of evolution, which is fourth density. This chapter is an attempt to give a brief overview of the process of this evolution or transformation and its implications on our lives and how we can cooperate with it to navigate these unprecedented times smoothly.

As per biologists, the history of the earth is broadly divided into four great eons from the formation of the planet some 4.5 billion years ago. The first one is called Hadean eon which consists of the first density stage of the planet and the second one is Archean eon which consists of the initial second-density stage when organic life of unicellular forms started to evolve. The third eon is called the Proterozoic eon which consists of mid-second-density when complex organisms with multicellular forms evolved and the fourth one is the Phanerozoic eon consisting of late second-density and third-density when more complex life forms including vertebrates evolved. Phanerozoic eon started half a billion years ago and is continuing. Most of the animals and plants that have ever existed on the planet appeared during this eon. Modern animals including humans evolved at the most recent period of this eon. Biologists estimate that 99% of all species that had ever lived on our planet have become extinct so far, which means what now remains is only 1%, a minuscule portion of all the species that evolved during this eon. This proves that many major transformations happened to the planet during this eon that caused the extinction of most of the species and also the evolution of many new species.

At each transformation most of the dominant species become extinct and scientists are still collecting their evidence. Similarly, human beings, the dominant species of the current time, are also scheduled for extinction when the planet transforms into its next level, which is fourth density. As per the evolutionary cycle, the core vibration of our planet's metaphysical body (soul) has already moved into the green-ray that corresponds to the fourth density. The physical body which we generally consider as the planet is currently undergoing transformation and is expected to reach complete vibrational alignment with the metaphysical fourth-density frequency in a couple of centuries from now. Currently, it is not possible to calculate the exact duration of this transition period and how long third-density life could continue during the transition as that is also dependent on our collective mentality and behavior.

Once the planetary transformation is complete, those souls who have attained the requisite frequency and vibrate in universal love are going to graduate to positive fourth density upon this planet and those who vibrate in utter selfishness will graduate to negative fourth density on another planet. Whereas, those who could not attain the requisite frequency, which is the case for the majority of humans, will have to repeat another cycle of third density life by reincarnating on another third-density planet.

The Mechanism

Sun caused the creation of the planets and supports them to evolve to higher dimensions until everything returns to itself at its black-hole stage and becomes one with the Infinite Creator or God. In our solar system, the highest individualized form of God is the Sun, which consciously supports and maintains everything within the solar system. The metaphysical aspect or role of the Sun is not apparent when we look at it just as a physical object that provides light.

Even from a strictly scientific or material point of view, this orb of fire, that we call our Sun, releasing energy to the entire solar system via the fusion process going on in it, is not only the physical center for all the planets in our solar system but also the basis of biological life. We all know plants produce food through the process of photosynthesis, and without plants, no animal can survive. And photosynthesis is not possible without the Sun emitting light or photons and without light neither plants nor animals can survive. Some religions have shown great reverence towards the Sun and have attributed to it the status of 'a god'. For example, in Hinduism, the Sun is called '*Surya Bhagwan*', which translates as 'Sun god'. Many other ancient religions also considered Sun as a god.

The present evolutionary stage is comparable to that of the Cambrian period, more popularly known as Cambrian Explosion, at the beginning of the current eon, during which complex life forms of second-density emerged in a relatively short period. As with every transformation, Sun caused major changes to the planet during that period leading to favorable environmental and geological conditions for rapid evolution and diversification of the simple life forms that existed until then.

Similarly, the transformation that we are currently undergoing is also induced by our Sun. Metaphysically, it sends fourth-density vibrations and helps third-density vibrations of individual and planetary souls to evolve to fourth density. Physically it sends huge bursts of energy, that are called solar flares and coronal mass ejections (CMEs), and engulf the planet with an unimaginable quantity of radiation, ions, and electrons to aid in changing the vibratory rate and nature of the physical planet and everything on it. The present heightened solar activity or changes in the solar weather is a step by step process through which the planetary vibration will be transformed into fourth density. Over a period of time this increasing solar activity will transform the entire planet and its inhabitants to a higher vibration. Anything that does not or cannot change or cope with the higher vibrations will cease to exist.

Given below is a pictorial representation of how the Sun sends focused radiation towards planet Earth:

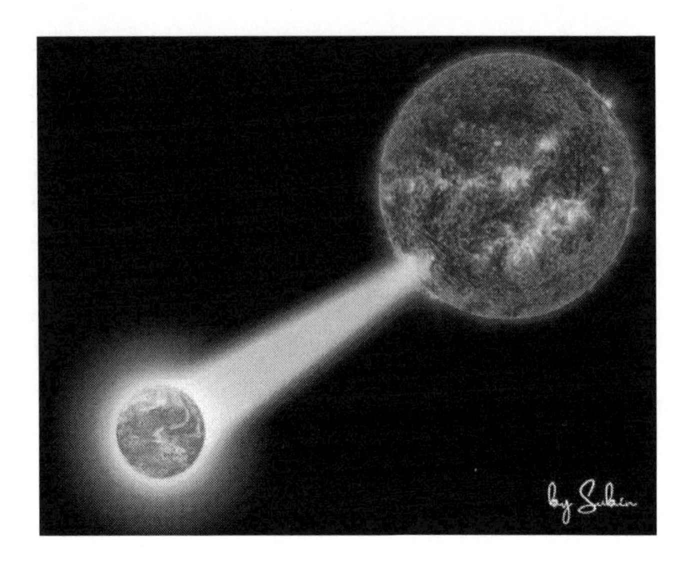

There are multiple changes happening simultaneously consequent to the cleansing and balancing that our planet undergoes while transforming itself to higher frequencies of fourth-density vibrations. Changes are slow initially and progress rapidly in an ascending order, with a few catastrophic events from time to time. By the end of the transition period, humans and most of the second-density species will be extinct from the planet and some new second-density species, that can survive in harsh climate, may emerge. Like we currently coexist with first and second density souls on this planet, once the transition is complete, the earth will host fourth (above the ground), second and first density souls, until it again becomes habitable for humans and normal plants and animals.

If we use a computer analogy, then our earth is being upgraded to the next higher version of the operating system and the hardware is being suitably modified to the extent it can support the new operating system. We, like the individual applications running on a computer, must ensure that we also become compatible with this new operating environment so that we can continue to function on this planet as fourth-density beings.

Impact of planetary changes

Solar flares send out light and radiation at a relatively smaller scale whereas the CMEs send billions of tons of solar material that includes condensed forms of light (like plasma) and several radiations (like X-rays and gamma rays that are harmful to third-density life). As the present atmosphere of earth allows only a fraction of these materials to reach the surface, there is no instant harm done, especially to human beings who are more susceptible to such harmful radiation. However, this ongoing activity reduces the thickness and transforms the composition/ structure of the atmosphere and makes it possible for more and more solar material to pass through it, causing a major change in our environment. We know that the climate of a planet is mainly dependent on two factors - its distance from the Sun and the nature and condition of its atmosphere.

Two major issues the world is currently struggling with are the accelerating environmental and social changes. We can already see changes like melting glaciers, shifting axis, changes in weather patterns (extreme climatic conditions), rising sea levels (mainly due to global warming), increase in frequency and intensity of rains, floods, storms, earthquakes, etc. Many are experiencing physical and mental discomfort and changes in food and sleep habits. Everything is changing so fast that many are not able to cope with it. There is a lot of confusion and many people are increasingly resorting to violence and extreme measures like suicide. There is an increase in conflicts among countries and the possibility of war, which is a serious risk because of weapons of mass destruction.

Organisms are prone to health issues when they are unable to align with the higher vibrations of fourth density. In the case of humans, this leads to mental issues also. Most of those newfound diseases (both physical and mental) that now plague humanity are consequent to the inability to adapt to the accelerating fourth-density vibrations and consequent planetary changes like depletion of ozone in the stratosphere leading to increased ultraviolet radiation and climate changes, compounded by the toxins in the soil, air, water, food, medicines, and an imbalanced mind along with thousands of different radiations emitted by power and communication systems. Pollution together with planetary changes is also causing the extinction of certain plants and animals which could adversely affect the food security of the planet.

Despite challenges, the ongoing transformation has brought with it many opportunities as well. The positive effect of the transformation is that there is an explosion in knowledge, development of technology and communication, rapid political and economic changes, and more people becoming rational and seeking freedom and expression. With the advent of internet and globalization, physical borders have lost their original meaning and the world is increasingly becoming connected and one. Rigid and corrupt social/ political systems are fading into history though every attempt is being made to maintain their status quo. Also, the new higher vibrations of energy caused and continue to cause changes in the pattern and vibratory rate of the physical/chemical bodies of living organisms and is unlocking latent capabilities of the DNA.

Many are being awakened to the positive fourth-density vibrations that are of truth, love, peace, happiness, freedom, and oneness. It has made the survival of dictators and rigid political and religious systems difficult. Value systems have changed for the better; more and more people have begun asserting their right to have a dignified life without the interference of corruption and injustice. As we move further into the higher vibrations of fourth density, more people will progressively express the qualities of higher nature.

Spirituality is being redefined and people are increasingly becoming disillusioned with traditional religions and seeking the truth beyond their set doctrines and systems. Internet and social media have created opportunities for learning, communication, and self-expression like never before. Higher vibrations together with technology have made it difficult for manipulators to keep the masses in perpetual darkness. Many have already realized, and more are being realized that essentially humanity is one and problem of one country or a community or an individual is a problem to all, and we will be happy and peaceful to the extent our fellow beings are happy and peaceful.

We can see that the feedback mechanism (the Law of Action or Karma) has become faster and one does not have to wait long for results. Evil deeds get punished quickly, and good deeds get rewarded swiftly. Any misuse/ abuse will lead to problems for the doer faster than in earlier times. The higher frequency of the soul and its environment causes thoughts to become things at a speed never seen before. This is the reason what we focus (intentionally or unintentionally) soon becomes a reality in our life. The period of waiting has come down and we can create/ manifest what we choose faster.

How to make the transition smooth

Planetary changes are the biggest challenge humanity will be facing going forward. The ill-effects of planetary changes are compounded when the majority of people on the planet live in disharmony and pollute the planet as that will delay/ obstruct the healing/ balancing process leading to further complications and suffering. Though we may not have a direct role in the transformation of the planet, we are responsible for polluting the planet both physically and mentally and thus making the transition process difficult.

When we know who we are, the evolutionary nature of life, and the oneness of everything, then we can lead a peaceful, purposeful, and happy life and make meaningful progress in our soul evolution or spiritual development. Spiritual development helps us to live in alignment with the fourth-density vibrations of the planet. Those who can evolve with the planet will be least affected by the ill effects of the planetary changes. To make this transition faster and smoother for the planet as well as for each one of us, we should know the truth about this transformation and consciously help the planet and each other with an open mind. When we help others then we will also be helped because of the Law of Action. The difference between an ignorant person and a spiritual person is that the ignorant person is a slave to his mind and environment, whereas a spiritual person has achieved a certain level of control over his mind and the environment and hence can manage/ change it.

The current environment of our planet is more conducive than ever for people who can work on higher *chakras* (heart onwards) to evolve to higher levels of awareness because of the higher vibrations present in the metaphysical and physical environment around us. What we have at hand is a golden opportunity and if we choose to use it wisely then we will be able to bring wonderful changes in all areas of our lives individually and collectively. When the majority of humans develop spiritually and live naturally - in harmony with each other and Nature - then our planet will also be supported in its efforts of achieving proper alignment and transformation to fourth density.

Those who choose to live in the light of truth can ascend with the planet and those who choose to live in the darkness of ignorance are going to suffer because of their inability to handle higher vibrations. Understanding and accepting our self and the essential unity or oneness of everything is the key to our graduation to fourth-density positive. When our consciousness shifts from an egocentric one to an open and universal one, where we consider All as One, then we will know we are ready for graduation.

****

Epilogue

Science and Spirituality are two sides of the same coin and we cannot separate one from the other. Science is trying to find the truth objectively or physically whereas spirituality finds it subjectively or metaphysically. Either one, if used alone, can only reveal one side of the truth and lead to an imbalance. Now the time has come for the true integration of science and spirituality so that there can be a revelation of the whole truth that can bring sustainable development and peace on planet earth. Einstein reckoned that we must change our way of research of the world to include broader and more multidimensional perspectives. No wonder Einstein was able to make such discoveries that were beyond the material capability of his time and helped humanity to make a quantum leap in Science and Technology.

To understand the relationship that exists between the physical and the metaphysical, one must conduct scientific analysis using both physical and mental capabilities in the light of available scientific data and spiritual truths revealed by all the great masters and genuine extra-terrestrial sources. This approach is holistic and will lead to freedom, peace, and happiness whereas lop-sided approaches lead to slavery and destruction, as we have on earth now.

There is nothing in the universe that can exist apart from the Creator. The only difference amongst creations at different levels of existence is the degree of awareness. Because we are made in the image and likeness of God, we have the same creative abilities as God at an individual level and can exercise this power proportionate to the degree of awareness we have. Each one of us has the freedom to create our future as we choose within the limitation of the natural laws.

Spirituality is not about leading a religious life though each religion has some elements of spirituality in it. Mostly religion enslaves the individual and keeps the society divided whereas spirituality makes the individual free and unites the society. A religious person identifies himself with a certain group of people whereas a spiritual person identifies himself with All Creation. We need to have discernment and we have to discard those things that do not support our spiritual development and focus our attention on higher truths that can take us forward.

Keep in mind that there was a time in recent history when people believed that earth was flat with humans living on the upper surface and below it was hell where evil spirits resided, and the sky above was where gods/ good spirits resided. As per them, the Sun and Moon were created by the gods to provide light to earth and stars were small light bodies used for decorating the night sky. Till recently, torturing and killing fellow humans for entertainment and religious purposes was common and the majority did not find anything wrong with it (this type of thinking is still prevalent in certain sections of the society). It is also true that one or two awakened individuals in each generation challenged those false beliefs and tried to tell the truth and in turn were persecuted and even killed. Look at the history and see which great teacher did not suffer for telling the truth in advance.

Mind is the greatest tool that man has at his disposal. What we are today is the result of our past thoughts and what we will be tomorrow is dependent on our today's thoughts. This is true both at an individual level and at the collective level. Our thoughts attract to us corresponding circumstances and environments which determines whether we will be happy or miserable, successful or unsuccessful, healthy, or unhealthy.

I believe that the time has come for human beings on planet earth to take control and responsibility for their own lives and live in cooperation with other beings and Nature and evolve consciously. This is the only sustainable solution before humanity now, which can save us from the self-destructive and unipolar approach to 'development' and allow us to consciously cooperate with the evolution of the planet to the next level of its existence.

This theory of life would stand up to both scientific and logical scrutiny. It is my sincere hope that the truths expounded in this book will help you to understand the nature and mechanism of life and equip you to lead an abundant and blissful life, and consciously evolve to higher levels of awareness.

Be a sincere seeker and not a blind follower. Do your thinking – examine and question everything and confirm the answer for yourself. Keep in mind that there are always more truths to know, and we will know more as we evolve. The more truths we know, the more will be our capacity to know higher truths.

I thank you very much for taking out the time to read this book and wish you a speedy and happy conscious spiritual evolution.

If you have any queries about these concepts or spiritual matters in general, please let me know through facebook or email. I will try my best to respond to them at the earliest.

Thomas Vazhakunnathu

****

May I ask you for a small favor?

I once again sincerely thank you for taking out time to read this book. You could have chosen any other book, but you took mine, and I totally appreciate this.

Hope you got at least a few new insights that will have a positive impact on your life.

This book may need multiple readings to get better clarity on certain subjects and you will definitely gain more value with each reading.

May I now ask for 30 seconds more of your valuable time?

I'd love if you could leave a review about the book. Reviews may not matter to famous authors; but they're a tremendous help for first time authors like me, who don't have much following. They help me to grow my readership by encouraging potential readers to take a chance on this book.

To put it straight– **reviews are the life blood for any author.**

It will just take less than a minute of your precious time but will immensely help me to reach out to more people. I am mindful of the fact that you may feel better equipped to share a review on a subsequent perusal. However, you can just give a rating and general comment for now.

Thanks for your support to my work. And I'd love to see your review.

Blessings

Thomas Vazhakunnathu

About the Author

Thomas Vazhakunnathu is a spiritual scientist, teacher, and author. His conscious spiritual journey started at the age of 23 in the year 1994 when he had a profound spiritual experience that changed the course of his life. Since then he devoted his life towards seeking and living the fundamental truths of life. By the year 2004, he understood that there is no religion or path that can claim exclusivity to truth or salvation. Further, he understood the oneness of everything, and that there is 'truth' in everything, albeit with varying degrees, and knowing these truths, particularly the fundamental truths, can free us from many limitations and sufferings and lead to peace and happiness in an ascending order, and ultimately to a blissful existence.

As a conscious observer of life, he enjoys spending most of his time investigating, understanding, integrating, and living the fundamental truths of life. His ultimate objective for the present life is to awaken and transform as many people as possible and contribute to peace, health, and harmony on planet Earth. He is open to working with anyone who shares this objective. He places more emphasis on living the truth or walking the talk than anything else as that is the only way one can move up in the scale of life. For him, there is nothing more important than living a blissful and purposeful life with an ever-expanding consciousness. His formal education includes a masters in English Literature and in Business Administration.

He can be reached online at:

Website: http://www.thomasvazhakunnathu.com
Facebook:
http://www.facebook.com/tvazhakunnathu
Twitter:
http://twitter.com/tvazhakunnathu
Instagram:
http://instagram.com/tvazhakunnathu
Email:
thomasvazhakunnathu@gmail.com

****

Acknowledgement

I acknowledge that the truths discussed in this book could not have been brought to light without the help of the ancient masters, the evolved souls living today, and the guides from the metaphysical world.

I wish to express my sincere appreciation of the following major sources for helping me to spiritually develop, understand, and explain the truths contained in this book:

1) The Ra Material from L/L Research
2) Upanishads, Bhagavad Gita, and Yoga Sutras
3) The Bible
4) Hematology
5) Science of Mind
6) Quantum Mechanics
7) Astrophysics

Special thanks to my wife Ancy and sons Eddy and Ebin, without whose encouragement, support and cooperation I could not have completed this book. Also, I would like to specifically thank my friend Shobhit Bishnoi for editing the book, Subin Mathew, Klemen Vrankar and Alexander Andrews for the graphics, Ranjit Jose for creating the book cover, AFH members and Som Bathla for enabling me to publish. Apart from the above, there are many individuals and sources that influenced and supported me in this work, naming all of whom here would be practically difficult.

****

Disclaimer

This book contains several theories that are beyond the scientific analysis of our present times. However, the author does not claim omniscience or infallibility. Some of the suggestions may seem polemical and may not suit everyone. If some material is not making sense to you now, please do not make it a roadblock for assimilating that which is making sense. Take only what is open to you and work slowly towards higher levels.

The health information contained in this book is for educational and informational purposes only and is not intended as health or medical advice. Always consult a physician or other qualified professionals for medical advice.

Although every precaution has been taken in the preparation of this work, neither the author nor the publisher shall have any liability to any person concerning any loss or damage caused or alleged to be caused directly or indirectly by the information contained in this book.

****

Printed in Great Britain
by Amazon

67839525R00118